ALGAE

A Practical Approach

ALGAE

A Practical Approach

Dr. K. SIVAKUMAR

Associate Professor
Department of Botany
Annamalai University
Annamalainagar

Chennai   New Delhi   Tirunelveli

MJP
PUBLISHERS

ISBN 9798224393275   **MJP Publishers**

All rights reserved   No. 44, Nallathambi Street,
Printed and bound in India   Triplicane, Chennai 600 005

MJP   253   © Publishers, 2016

Publisher : C. Janarthanan

Dedicated To

AVUL PAKIR JAINULABDEEN ABDUL KALAM

Kalam's easy and friendly approach was a great asset in organizing the effort. He would effortlessly go to a technician in the workplace and learn first hand about his problems and provide him encouragement. No wonder the whole group working under him warmed to this approach and always extended their unstinted support and enthusiasm.

❧ PREFACE ❧

Laboratory studies constitute an integral and important aspect in the study of science subjects. Botany in particular is a subject heavily oriented on practical study. A proper comprehension of the subject matter is possible only in the laboratory. Laboratory manuals play an important role in helping the students to properly guide them in purpose of study, scope of study and details required for a practical study.

Practical manual of algae aims to provide the students all they need to know about the practical aspects such as morphological characters, internal structure if any and systematic identification. A number of labelled diagrams included in the text are meant to help the student's comprehension of the subject.

This manual "Algae – A Practical Approach" deals with all the basic requirements in the botany laboratory that a student should know. The working of compound microscope, the most important tool in the laboratory is explained in great detail in the introduction, along with various procedures of study such as section cutting, staining, mounting, etc., working principles of electron microscopy and material preparation for scanning and transmission electron microscopic studies. General characters and classification of algae with all essential descriptions and identification of representative members of various classes of algae have been provided in this manual.

With the increased awareness of importance of laboratory studies, this manual should serve as an important aid to help students familiarize themselves with laboratory methods in botany.

I am thankful to the authorities of Annamalai University for having granted permission to write and publishing this manual.

Finally, I am indebted to MJP publishers, Triplicane, Chennai for timely publication of this manual in such a neat and excellent manner.

Criticisms and suggestions from the actual users of the book are welcome.

Dr. K. SIVAKUMAR

Associate Professor
Department of Botany
Annamalai University
Annamalainagar

❧ FOREWORD ❧

Dr. K. Kathiresan, M.Sc., Ph.D., D.Sc.,

Professor and Director
Dean, Faculty of Marine Sciences
Annamalai University, Parangipettai
Tamil Nadu, India

Algae are unique photosynthetic organisms, supporting like "mother milk" to the food web of aquatic environment. The algae are ubiquitous, cosmopolitan and occurring in a range of habitats from lentic, aquatic to subaerial or aerial, xeric, snowy, pollute and unpolluted and are endowed with the capability to survive in hostile environment.

Microalgae are of great ecological importance and economic significance as a source of food, feed, fuel, fertilizers, fine chemicals and pharmaceuticals and their role in pollution abatement. To make use of more such organisms, it is essential first to identify their biodiversity in the naturally occurring populations and establish their germplasm for further characterization.

Bearing this in mind, the "Algae - A Practical Approach" has been prepared to provide the students what they need to know about the practical aspects of morphological characters, internal structure and systematic identification. A number of labelled diagrams provided will be helpful to the students to understand the subject.

There is ample information, available on various taxa of algae. This practical manual will be an excellent document for the benefit of students and teachers as well as useful for researchers interested in the biology of algae.

My hearty appreciation is due to Dr. K. Sivakumar, Associate Professor in botany for his sincere effort in bringing out this manual "Algae - A Practical Approach".

(Dr. K. KATHIRESAN)

Professor and Director
Dean, Faculty of Marine Sciences

CONTENTS

❧ 1 ❦

INTRODUCTION

The author, Dr.K.Sivakumar, started his research career during 1990-1992 in the Department of Botany, Annamalai University in the field of Algology, specializing on the morphology, anatomy and biochemistry of *Sargassum* sp., collected from the Pamban Coast. Then, he moved to the Centre for Advance Studies in Botany, University of Madras and pursued research under the guidance of Prof. R. Rengasamy, former Director of the Centre. Under his able guidance, he started the research work on the ultrastructure of red algae. Transmission electron microscopic studies were made on the vegetative thallus of the carrageenophyte of *Hypnea valentiae* (Turn.) Mont, collected from the Kurusadai Island, which revealed that the outer cell wall layer was more dense with electrons than the inner layer. The cortical region consisted of one or two layers of small and rectangular cells and had dense cytoplasm packed with plastids and starch grains. The subcortical region had vacuolated cells. Long plastids were localized at the periphery. Medullary region consisted of highly vacuolated large cells. Thin peripheral cytoplasm contained elongated and highly irregular plastids. Pit connections consisted of pit plug and pit core. An electron dense region was also observed near the margin of the plug core which in turn was surrounded by a cap membrane. The Golgi complex was in close proximity to the nucleus. Both smooth and rough endoplasmic reticulum were closely associated with the Floridian starch grains. Carposporogenesis and tetrasporogenesis were also studied. This work is first of its kind in India.

The author received from UGC a Major Research Project entitled "Biodiversity of fresh water microalgae from Veeranam lake, and associated rice field from Cuddalore district of Tamilnadu". In this project, phytoplankton samples were collected from the Veeranam lake and studied for a period of one year from March - 2005 to February - 2006. Physicochemical parameters of the lake water such as air temperature, water temperature, pH, salinity, dissolved oxygen, electrical conductivity and total dissolved solids were observed and their ranges were: 30.1-36.5°C, 29.0-34.4°C, 7.9-8.4, 1.2-2.5 mgl^{-1}, 7.6–9.2 µS and 2.5-5.2 mgl^{-1} respectively. Totally 160 species of phytoplankton belonging to different taxonomic groups were identified. Among these, 74 species belonged to Bacillariophyceae, 43 species to Chlorophyceae, 38 species to Cyanophyceae and 5 species to Euglenophyceae. The phytoplankton density was high (1705 cells l^{-1}) during the summer season and lower (760 cells l^{-1}) during the winter season. Bacillariophyceae formed the dominant group. Species Diversity Index (H'), Species Richness (SR) and Species Evenness (J') were also calculated.

Commonly occurring genera, *Oscillatoria* (Cyanophyceae), *Navicula* (Bacillariophyceae) and *Scenedesmus* (Chlorophyceae) were subjected to Energy Dispersive Spectroscopic analysis (EDS). They were found to accumulate different elements such as Zn, P, S, Ca, Mg, Fe, N, Si, Cl and Mn. Among these, *Oscillatoria* contained Zn, P, Mg, Ca, Mn, S and N. *Navicula* had Si, Zn, Mg, Cl, N, Fe and Ca. *Scenedesmus* possessed Ca, Mg, N, Fe, Cl, Zn, Si and Mn. These observations would help to determine the chemical dialogue between the cell structures and role of the elements in addition to the clue about the phytoplankton growth requirements.

Ecology and biodiversity of freshwater microalgae in Chidambaram, Cuddalore district of Tamilnadu with special reference to phycoremediation were investigated. From the study, it was apparent that the water quality was affected due to anthropogeanic activities coupled with seasonal phytoplanktonic dynamics. It could be inferred from the values of dissolved oxygen and total dissolved solids of Omakulam pond and Palaman River that these water bodies are facing pollution threat. Throughout the study period, pollution-tolerant algae belonging to Bacillariophyceae and Euglenophyceae were dominant indicating the presence of higher organic load.

Regarding the relationship between the seasonal variation and phytoplankton dynamics, a significant increase was noticed during the summer season, which were linked to the availability of water minerals, particularly phosphate-phosphorus. Bacillariophyceae was dominant in the dry season while Cyanophyceae was dominant throughout the year followed by Chlorophyceae, which increased in the wet season. Therefore, Omakulam pond and Palaman River could be "Mesotrophic pond" during the dry season and "Eutrophic" during the wet season. This would mean that the water quality is under threat, due to the mineral enrichment. So, it is a matter of concern and awareness should be created on the impacts of land use in the reservoirs and water shed areas.

Although certain abatement technologies can be developed to clean the polluted environment, there is an urgent need to educate the people about the ill effects of pollution and to curb pollution in our pristine environment because "prevention is better than cure and prevention is the best way to cure rather than abatement". Further, all the land use activities around the reservoir need to be managed appropriately and sustainably, in order to protect the health of the ecosystem, from adverse conditioning including algal blooms.

To study the role of cyanobacteria in heavy metal removal from effluents, samples were collected from the tannery and heavy metal analysis was done. Among the heavy metals studied, Cr and Zn showed higher contents. These two metals were selected for further experimental works. The cyanobacterium, *Oscillatoria terebriformis* was isolated from the place where the effluent was collected. The effluent samples were treated with *O. terebriformis* both under sterilized and unsterilized conditions. To know the effect of effluent (containing heavy metals) on the cyanobacterium, biochemical studies were also conducted in the effluent treated alga along with control. Results indicated that the effluents with the heavy metals could be treated with the alga, as the alga helped in the removal of heavy metals and in producing a variety of value added products from the cyanobacterial biomass.

Cyanobacteria are abundantly present in the rice fields and are important in maintaining the fertility of rice fields through nitrogen fixation and photosynthesis. Hence, the cyanobacteria isolated from the paddy fields of Vallampadugai and Keerapalayam, Cuddalore District, were maintained in BG11 medium for 10 days. The ultrastructure of cell wall and the biochemical and pigment analyses of the isolated organisms was carried out and the results observed. Scanning Electron Microscopic (SEM), Transmission Electron Microscopic (TEM) and biochemical features (total carbohydrates, total lipids, total proteins, total free amino acids, and photosynthetic accessory pigments such as chlorophyll *a*, phycocyanin, phycoerythrin and allophycocyanin) were observed in *Anabaena* sp. and *Oscillatoria* sp.

Further, effect of cyanobacteria as biofertilizer on the growth and yield attributes of rice (ADT38) was evaluated. Growth parameters such as shoot and root lengths and yield attributes like plant height, number of tillers, number of panicles and grain weight were observed. It was found that the application of cyanobacteria enhanced the growth and yield parameters of rice as compared to control and improved the soil properties.

During the Post-Doctoral Fellowship period in Chungum National University, Dajean, South Korea, (October 2007 to March 2008), studies were made on the mature vegetative thalli of *Gelidium elegans,* using scanning and transmission electron microscopy. Structural evidences suggested a storage rather than a secretary function for the cells. Rhizoidal cell walls were relatively thin, which in turn would aid in the transfer of metabolites of the secretary/storage structures. These features of the rhizoidal cells gave essential clues to the production and secretary storage functions of the halogenated metabolites in *G. elegans,* offering new insights into the possible mechanism for their release.

Hypnea musciformis (red alga) was investigated for the richness of its bioactive compounds. Its ethanolic extract was subjected to GC-MS analysis and it revealed 26 chemical constituents including bioactive major constituents like n-hexadecanoic acid- tetradecanoic acid, oleic acid-9-octadecenoic acid-6-octadecenoic acid, hexadecanoic acid-ethyl ester-ethyl tridecanoate and octadecanoic acid. Further biochemical and mineral analyses revealed the presence of phenols, total carbohydrate, total protein, fat, moisture content, ash, vitamins A, C and E and lipid soluble and water soluble antioxidants. Chemical elements such as Ca, Mg, Na, K and Fe were in highest proportion.

Another study revealed that *Sargassum wightii* (brown alga) had antihyperglycemic, antioxidant and antihyperlipidemic activities in STZ-induced male albino wistar rats. The alga also showed protective effect against kidney, liver and pancreatic injuries, associated with diabetes. Methanolic extract of *S.wightii* increased the insulin secretion, enhanced the insulin stimulated glucose uptake by glucose transporters and modulated the lipid and carbohydrate metabolism by up-regulating peroxisome proliferator-activated receptors. In addition, methanolic extract enhanced the secretion of incretin hormones in the intestine and the released incretin hormones promoted insulin secretion which, in turn, exerted glycemic control, altered lipid profile and maintained antioxidant status to a near normal level. The biochemical studies of tested animals were corroborated by histopathological studies and a small piece of tissue contained 9-12 different elements, which were recorded from the individual intracellular portion of liver using Scanning Electron Microscopy with Energy Dispersive Spectroscopic analysis. This is perhaps the first report from India.

Chemopreventive and antioxidant effects of *S. wightii* and *H. musciformis* on 7, 12 -dimethylbenz(a)anthracene (DMBA) induced buccal pouch carcinogenesis in hamster were studied. Oral squamous cell carcinoma was induced in buccal pouches of male golden Syrian hamsters by painting with 0.5% in lipid paraffin three times per week for 14 weeks and there was 100% tumor formation in DMBA painted hamsters. Oral administration of methanolic extracts of *S. wightii* and *H. musciformis* (200/300 mg/kg b.wt) prevented the tumor formation. A decrease in the levels of lipid peroxidation byproducts was observed, which revealed the enhancement of the defense mechanism of the antioxidants in DMBA painted hamsters. The methanolic extracts of *S. wightii* and *H. musciformis* might probably act by altering the status of lipid peroxidation and antioxidants in DMBA painted hamsters.

To know the effect of the Seaweed Liquid Fertilizer (SLF) of *Ulva reticulata*, on the biochemical characteristics of *Vigna mungo* as well as its leaf morphometric characters such as epidermal and stomata cell variation and distribution of minerals in the leaf, experiments were conducted. Seeds of *V. mungo* were sown in the soil and SLF was added to the soil bed in five different concentrations (1%, 2%, 4%, 6% and 8% w/v) separately. SLF was also applied as a foliar spray; the treated plants showed maximum growth in 2% of SLF among the various experimental concentrations as well as control. Biochemical profiles like chlorophyll *a* and *b*, protein, sugar and starch were also found to be higher at 2%. A significant increase in the number of epidermal and stomatal cells was observed in 2% SLF treated plant leaves. Whereas, at the higher concentrations of the SLF (4%, 6% and 8%), values of all the parameters were significantly lower than in the control group. Further, the leaf of 2% SLF treated *V. mungo* was subjected to Scanning Electron Microscopy with Energy Dispersive Spectroscopic analysis and it revealed the presence of ten elements in the following order: Ca > P> N > Na> K > Mg > Mn > S > Fe >Zn in the treated plant and Ca > N > P > Na > Mg > Mn > K > Zn > S

> Fe in the control plant. Thus, study revealed that SLF of *U. reticulata* could be used as foliar spray at a lower concentration of 2% to maximize the growth and yield of *V. mungo*. This study would pave way for organic farming and it would be cost-effective and eco-friendly for promoting sustainable agriculture.

Biosynthesis of silver nanoparticles using *H. musciformis* along with photocatalytic degradation of methyl orange dye was another aspect of research. Nanoparticles of silver were formed by the reduction of silver nitrate to aqueous silver metal ions during the exposure to the extract of the red alga, *H. musciformis*. Photocatalytic degradation of methyl orange was measured spectrophotometrically by using silver as nanocatalyst under the visible light illumination. The results revealed that the biosynthesized silver nanoparticles using *H. musciformis* were found to be effective in degrading the methyl orange. Optical properties of the obtained silver nanoparticles were characterized by UV-visible absorption and room temperature photoluminescence. X-ray diffraction results showed that the synthesized silver nanoparticles were in the cubic phase. Existence of functional groups was identified using the Fourier Transform Infrared Spectroscopy. Morphology and size of the synthesized particles were studied using the atomic force microscopic measurements.

Thus, the excellent potentials of both the freshwater and marine algae can be harnessed for human and environmental benefits. So, research institutes, Government, agencies and private entrepreneurs with scientific and technical knowledge and with marketing expertise should come forward as the value added algal products are having pharmaceutical, nutraceutical and cosmaceutical properties.

The author has a rich silver jubilee experience of twenty-five years in the field of algal research. This experience has resulted in his compilation of the present book with respect to the excellent potentials of both the freshwater and marine algae and their human and environmental benefits which would be of immense use to the college-going and research students.

2

MICROSCOPY AND STAINING

The study of science subjects is carried out at two levels, viz., theoretical and practical. Theoretical study is carried out in lecture classes when the students are imparted all the details regarding a particular topic. This has to be supplemented by practical or laboratory studies, when the students come into active contact with the objects about which they would have heard in lecture classes. Therefore it may be safely said that laboratory studies form an integral part of a comprehensive study of any topic in science subjects.

In the study of algae, laboratory assumes a special significance as the students acquire a meaningful and first-hand knowledge about the algae which they study in lecture classes. In fact, as far as botany is considered, any study of algae without being supplemented by laboratory observation is futile. It is only in laboratory study that the student can dissect, observe and analyze the various features of the algae and it is only a study of this type that makes the student acquire a meaningful knowledge about the micro and macroalgae. Besides observation and study in the laboratory, algae are also to be studied in field to obtain an account of their habitat, habit, association and such other details.

In this chapter an attempt is made to explain some of the essentials that a student should understand before he begins a practical work.

Cleanliness: This is the most important point that a student should keep in mind while in the laboratory. The working table and the surroundings should be kept dry and spotlessly clean. All the requirements needed by the students such as pewter dishes, slides, reagent bottles, stains, etc., must be properly labelled and arranged neatly. The microscope and particularly the optical parts must be cleaned with a white lint free cloth. After the completion of the assigned work, the student has to take care to clean the place, remove any debris of the algae material and put it into the dustbin.

The students should work in silence and independently unless the assigned experiment is to be carried out by a group. When a particular work is assigned to a student, he must carefully plan and divide the available time for examination, observation and recording of the results. Whenever a doubt arises, it is always better to consult the teacher in charge, than a classmate. It is only by a proper and systematic work, that a student can derive maximum benefit from the practical classes.

LABORATORY REQUIREMENTS

The department provides certain amenities, instruments, etc., for each student for purpose of practical work. But a student also should carry some of the requirements in order to be effective in a practical class. One cardinal principle that every student should remember in practical class is to have one's own requirements and not to depend on anyone. Never lend or borrow any requirements. Borrowing or lending a pencil, needle or blade will disturb the work of both.

In general every student should individually possess the following requirements

1. *Two sharp pointed needles* mounted on stainless steel or plastic handle. This is essential for the purpose of separating and teasing the microscopic material.

2. *A pair of forceps*: This is essential to remove the material either from the bottle or from pewter dish.

3. *A sharp* properly honed razor to cut sections of the algal material.

4. A new safety razor blade. This will help either to cut section or to trim the material.

5. Two fine hair paint brushes. This will be used in lifting the sections of algae from the dish on to the slide.

6. Two sharpened pencils (H, Hb or 2b).

7. A pencil eraser.

8. A pencil sharpener.

9. Glass droppers.

10. A piece of thin, white, lint free cloth to clean the optical parts of the microscope.

11. An observation book to note down immediately the points of interest.

12. A practical record.

13. Glass slides and cover slips (if they are not provided by the department).

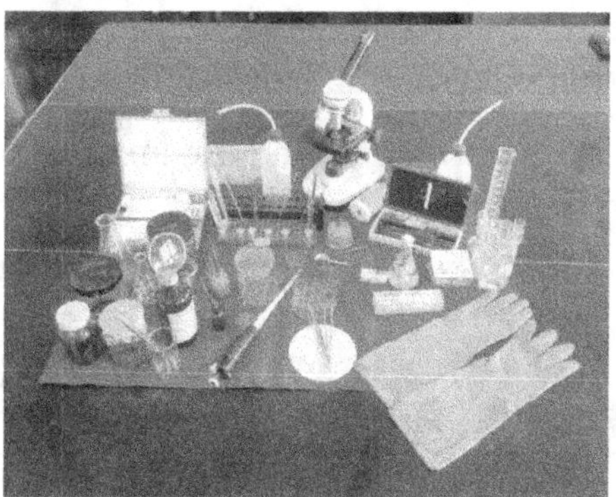

Laboratory accessory material

SCHEME OF PRACTICAL WORK

Practical work in the laboratory will be most useful and effective if student keeps in mind the following:

1. Maintain silence during working. All the requirements for the practical work must be absolutely clean.

2. Every student should work independently, unless the experiment involves the participation of others.

3.   Before beginning the practical work, listen to the instructions of the teacher attentively and work accordingly.

4.   Follow a systematic procedure to study.

5.   If the work involved the morphological study of an organism, follow the procedure given below:

   a.   Study of external characters - for example whether an algae is microscopic or macroscopic, unicellular or multicellular, filamentous or colonial, etc.

   b.   Study of internal structures like details of cell, tissues, etc. If the material requires sectioning, sections are to be cut and internal details should be observed.

   c.   Study the reproductive details like asexual and sexual reproductive organs.

6.   Mount the material after properly teasing and separating the filaments, etc. Stain the material wherever necessary and cover the preparation with a cover glass.

7.   Observe the slide carefully under the microscope, first under low power and then high power.

8.   Make sure that you have observed all the details as instructed by the teacher.

After all the observations are over, record your findings, structural details, etc., in the practical record in the form of diagrams. Draw proportionate diagrams in the proper places on the drawing sheet. Diagrams should never be drawn to the very edge of the drawing sheet. Right suitable explanatory notes pertaining to the diagrams should be given on the interleaf.

Laboratory Equipments

In addition to the individual requirements carried by the student, the department generally provides the following equipments/instruments, etc., to the students. Sometimes these equipments may be given to each student.

These are:

1.   Student's compound microscope.

2.   Dissection microscope.

3.   Watch glasses.

4.   Binocular research microscope.

5.   Cavity blocks.

6.   Slides and cover glasses.

7.   Pewter dishes.

8.   Enamel tray.

9.   Polythene/glass wash bottle.

10. Spirit lamps.

11. Glass marking pencil.

12. Microtome.

13. Staining bottles.

14. Stone for honing the razor.

15. Chemicals and stains.

16. Culture media (to be prepared by students).

In the list of requirements mentioned above, we will discuss some in detail.

MICROSCOPE

Of all the instruments used by students in the botany laboratory, the one that is most indispensable to the students of biology is the 'Microscope'. It is no exaggeration to say that no biological study in the laboratory is possible without the microscope. The function of the microscope is to enlarge the small objects, so that they become visible to the human eye. In reality the magnification power of the microscope lies in its resolving capacity between one particle and another. A microscope distinguishes one particle from another which is situated very closely. A human eye cannot resolve particles that are lying closer than 0.025 mm. In its function, a microscope can be compared to a human eye, but much more effective than that. There are different types of microscopes, but all of them are based on the same principle with variations in their construction. Some of the types of microscopes used in the laboratory are:

1.   Compound microscope

2.   Binocular Research microscope

3.   Dissection microscope

4.   Stereoscopic dissection microscope

5.   Phase contrast microscope

6.   Polarising microscope

7.   Fluorescence microscope

8.   Confocal laser scanning microscope

9.   X-ray microscope

10.   Scanning acoustic microscope (SAM)

11.   Scanning helium ion microscope (SHIM or HeIM)

12.   Scanning probe microscopes

13.   Scanning electron microscope

14.   Transmission electron microscope

15.   Scanning transmission electron microscope (STEM)

16.   High voltage electron microscope (HV-EM)

17.   Scanning tunnelling microscope (STM)

18.   Moist environment ambient temperature scanning electron micro scope (MEATSEM)

Of the eighteen types of microscopes mentioned above, the compound microscope and the dissection microscope are the ones most frequently used by students. The electron microscope is very costly and a very highly sophisticated instrument. In this book, the compound microscope, the dissection microscope, scanning and transmission electron microscope are explained in great detail as the students should be familiar with them.

Compound Microscope

This microscope has a strong 'U' shaped or 'V' shaped base which is also called the horse shoe shaped base. Attached to the base is a bent limb. This limb can be tilted in any direction as it is attached to the base with an inclination joint. Attached to the limb are a stage and a body tube. The body tube is attached to the limb with the help of a rack so that it can move up and down. To help in the movement of the tube during focusing, there are 2 knobs, viz., coarse adjustment knob and fine adjustment knob.

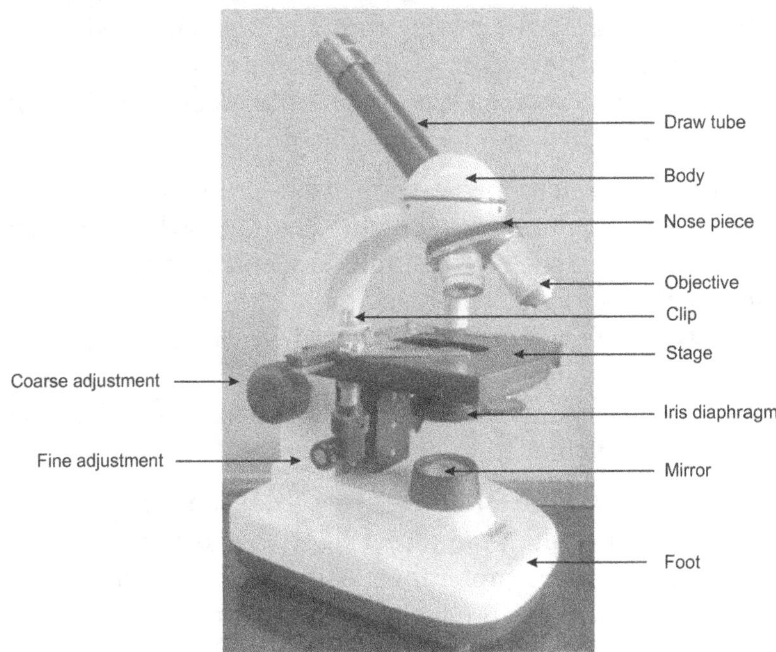

A Compound Microscope

The body tube at its top end consists of the ocular or eye piece lens. The eye pieces usually have three magnifications 5x, 10x, and 15x, and can be used to get a low or high magnification. Attached to the lower end of the tube is a revolving nose piece with three objective lenses low (10x), medium (40x) and high (60x). Sometimes there will be a fourth objective of the magnification of 100x. This is also called the oil immersion lens (while viewing through an oil immersion lens a drop of oil-liquid paraffin has to be put on the viewing slide).

Below the objectives and attached to the limb is the stage where slides are kept for viewing. The stage may be circular, square or rectangular in shape. In order to keep the slides properly placed on the stage, there are two stage clips. In some of the costly microscopes, the stage has movable sliding system operated by knobs to move the slides. This is called the mechanical stage. The mechanical stage permits the movement of the slide backwards or forwards, left or right with the turning of the knobs provided for the purpose. A hole is provided in the stage to receive the light from below and illuminate the object. Attached to the limb, below the stage is a condenser (fixed or movable up and down) lens. Below the condenser lens is an iris diaphragm to regulate the light passing through. Iris diaphragm is a metallic disc with an aperture. The aperture can be narrowed or widened at the turn of a lever. The condenser which is a system of lens condenses the light rays passing through and converges them on the object.

The source of illumination in a student's microscope is a mirror which reflects the natural light or electric light. The mirror has two faces plain and concave, and is attached to the limb. The mirror is adjustable and can turn to the direction of light. When day light is the source of illumination, the concave side of the mirror is usually used to converge the rays of light. In some of the costly microscopes, instead of the mirror there is a built-in illumination. This is a tube with a lamp (9v or 12v) attached to a transformer, which can be plugged to any AC source.

Operation of Microscope Microscope should be placed in a well illuminated area but avoiding the direct light as it is harmful to the eyes. Generally northern light is preferable. When built-in illumination is employed, it is always better to use a blue or green fitter to subdue the light. The light has to be first focused on the condenser by adjusting the mirror towards the source of light. The amount of light entering the condenser may be regulated by closing or opening the iris diaphragm. The slide to be viewed is kept on the stage and fixed securely with the help of the stage clips. The object is first observed with a low power objective and then turned on to medium or high power, the specimen or section has to be covered with a cover slip. The slide should be observed with both the eyes open even while observing with monocular microscope.

Converging Lens or Mirror

- A lens or mirror that can refract or reflect a parallel beam of light so that it converges at a point.

- Such a mirror is concave

- A converging lens is thicker at its centre than at its edges (i.e., it is biconvex, plano-convex or convexo-concave).

Diverging Lens or Mirror

- A lens or mirror that can refract or reflect a parallel beam of light into a diverging beam.

- A diverging lens is predominantly concave

- A diverging mirror is convex.

Care of the Microscope

1. Clean all the parts of the microscope before and after use. The lenses and the mirror are to be cleaned with a lint free cloth or tissue paper.

2. Do not pull the microscope to the very edge of the table.

3. While lifting the microscope, use both hands.

4. The slide to be observed must be clean and dry (especially the under surface of the slide).

5. Do not remove the eye pieces or objectives from the microscope.

6. The microscope should be covered when not in use.

Binocular Research Microscope

This is like the ordinary compound microscope except that, it has binocular head for providing a natural stereoscopic view.

Dissection Microscope

This is a very simple microscope with only one set of lenses. Microscopes of this type will not give a very high magnification, but are very useful in dissecting flowers, plant parts, etc.

The microscope has a 'V' shaped base and a limb. The stage is made up of a transparent glass plate. Below the stage is a mirror to focus the light. Attached to the limb is a mechanical arm which can be moved up and down with the help of a rack and pinion movement. The arm can also be moved horizontally. Attached to the free end of the arm is a holder to accommodate the lens (10x, 15x, etc.).

Operation   The specimen to be observed/dissected is placed on the stage and located with the help of lens. Floral parts or others may be then dissected with the help of two needles, while still observing the specimen.

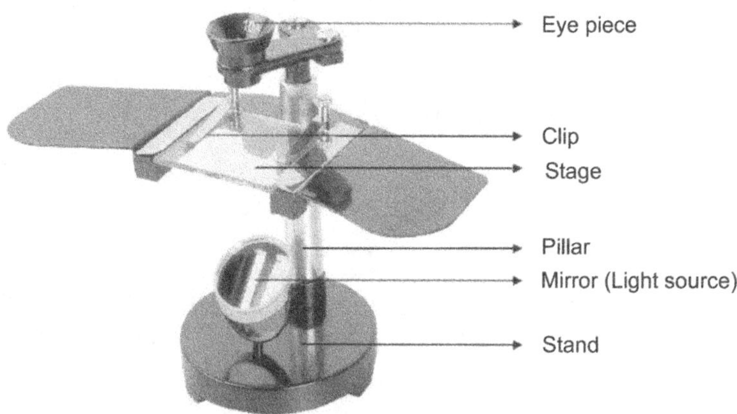

Stereoscopic Binocular Dissection Microscope

This is a combination of compound microscope and dissection microscope. While it has the higher magnification power of a compound microscope with two sets of lenses, it has retained the versatility and the manoeuvourablity of a dissection microscope.

Phase contrast and electron microscope are not normally used by students.

STAINS

These are colouring fluids. The specimens to be observed (algal or fungal material or sections of plant parts) would be generally decolourised due to preservation. These are to be coloured in order to study them clearly. The following are the stains that are generally used in the laboratory.

(1)   **Safranin**   This is the most commonly used stain. It stains nucleus and lignified cell walls.

Preparation

Safranin   1 gm

95% ethyl alcohol   50 mL

Distilled water   50 mL

An aqueous stain may also be prepared.

Safranin   18 mL

Distilled water   100 mL

(2) Fast Green

Fast green   1 gm

90% ethyl alcohol   100 mL

This stain is used to colour parenchyma cells and cytoplasm. Fast green may also be prepared in clove oil (75 mL clove oil + 25 mL 100% alcohol).

(3) Cotton blue

Cotton blue   1 gm

Lactophenol   100 mL

(4) Haematoxylin

The dye is obtained from the heartwood of the tree *Haematoxylon campechianum*. There are two methods of preparation of the stain

These are:

i. Heidenhains Haematoxylin

Dissolve 0.5 or 1.0 gm of the powder in 100 cc of distilled water. Filter and store in a dark bottle away from light. The stain is ready for use after 4-5 days of ripening. Ripening is not necessary if the stain is prepared in 70% ethyl alcohol.

ii. Delafields Haematoxylin

Solution 1: Aqueous solution of ferric ammonium sulphate 100ml.

Solution 2: One gm dye dissolved in 6ml of absolute alcohol.

Mix solutions 1 and 2. To this mixture add 25 ml of glycerine and 25 ml of absolute alcohol. The solution should be allowed to stand until it becomes dark red.

(5) **Aniline blue**

Also called cotton blue or China blue. It is prepared as follows:

Aniline blue   1 gm

Ethyl alcohol (90%)   100ml

Distilled water

Hydrochloric acid 2drops

(6) **Gentian Violet (Crystal Violet)**

Gentian Violet   1 gm

90% Ethyl Alcohol   100 ml

(7) **Erythrosine**

Erythrosine   1 gm

90% Ethyl alcohol   100ml

Erythrosine may also be prepared in clove oil (Absolute alcohol 0.5 ml + 95 ml Clove oil).

(8) **Light green**

Light green   1 gm

Clove oil   95ml

Absolute alcohol   5ml

Light green may also be prepared without clove oil (in 100 ml absolute alcohol).

(9) **Acetocarmine: The stain is prepared as follows**

Add 1 gm of stain to 100 ml of boiling acetic acid (45%) or propionic acid. Cool and decant, a few drops of saturated ferric acetate solution is added and the mixture is cooled over ice. The solution is allowed to stand for about 12 hours. Filter and store in a dark coloured bottle under refrigeration.

Specificity of Stains

Usually specific stains are employed in cooling different plant parts. The stains are used singly or in combination. The most popular stain combinations are:

* Safranin fast green

* Haematoxylin Erythrosin

* Haematoxylin light green

The table gives the specificity of stains to different plant parts or cell organelles.

Structure	Stain
Cellulose cell wall	Aniline blue, Haematoxylin, Fast green, Light green
Cutinized cell wall	Gentian violet, Erythrosine, Safranin
Lignified cell wall	Gentian violet, Safranin
Suberised cell wall	Safranin
Cytopalsm	Fast green, light green, Erythrosine, Aniline blue
Chitin	Safranin
Callose	Aniline blue
Achromatic figure	Aniline blue, Erythrosine
Proteins	Safranin
Nucleus	Heamatoxylin, Safranin, Gentian violet
Chromonemata	Heamatoxylin
Chromosomes	Safranin
Plastids	Gentian violet, Haematoxylin
Mitochondria	Gentian violet
Flagella	Cotton blue

While stain combinations are employed to study pteridophytes, gymnosperms and angiosperms, single stains are used for vascular cryptogams.

Green or safranin for algae and bryophytes.

Cotton blue for fungi.

FIXING/PRESERVING FLUIDS

The plants or plant parts collected in nature are to be killed and preserved in the laboratory for the purpose of storage and study. In killing the specimen, the various life processes are stopped at that instant, so that after preparation for microscopic study various stages may be observed.

Primary fixation on the spot

Whenever a material is collected, it is better to kill it in the field, so that there will be minimum or no distortion. Alternatively the material collected in the plastic bags or bottles should be brought to the laboratory and killed/preserved. In preservation, the material must be completely immersed in the fluid. The following are some of the killing/preservative fluids generally used:

1. *Formalin acetic alcohol (F.A.A)*

    This is the most commonly used killing/preserving agent. It is popularly called standard preservative.

Ethyl alcohol	90ml
Glacial acetic acid	5ml
Formalin	5ml

2. ***Carony's Fluid***

    Ethyl alcohol   30ml

    Glacial acetic acid   30ml

    Chloroform   15ml

    Carony's fluid is widely used in killing root tips, flower buds, etc., for the purpose of cytological study.

3. ***Formalin - Propiono - Alcohol***

    Formalin   5ml

    Propionic acid   5ml

    Ethyl alcohol 70%   90ml

4. ***Randolph 's Modified Navashin Fluid***

    This consists of two solutions A and B.

    Solution A

Chromic acid	5.0 gm
Glacial acetic acid	50 ml
Distilled water	320 ml

    Solution B

Commercial formalin	200ml
Distilled water	175 ml
Saponin	3 gm

    Mix A and B in equal volumes before use.

MOUNTING MEDIA

These are chemicals used for the purpose of mounting (temporary/permanent) the specimens on the slide for microscopic observation. The following are some of the mounting media.

1.   **Acid water**

HCl	10-15 drops
Water	100 ml

2.   **Lactophenol**

Phenol	100 gms
Lactic acid	100 ml
Glycerine	100 ml
Distilled water	100 ml

Glycerine

Glycerine	15-20 drops
Formalin	1-2 drops
Distilled water	100ml

3.   **Canada Balsam**

This is a gum obtained from gymnosperms. This fluid is readily available in the market.

4.   **Euparol**

This is available in the ready-to-use form in the market.

Canada balsam and Euparol are used for permanent mounting, while the rest are used for temporary preparations.

STAINING RACK

This is a wooden rack used for holding the staining bottles.

DROPPING BOTTLES

This is a glass bottle having a narrow mouth fitted with a stopper. The stopper is grooved and has a beak. When the bottle is up-turned the stain/fluid comes out in drops.

SLIDES

These are readily available in the market in their the usual size (25mmx-75mm). The thickness is about 1 mm. Special types of slides called cavity slides are available for specific study.

COVER SLIPS (COVER GLASSES)

These are meant to cover the materials when the preparation is ready for observation. The coverslips may be round, square or rectangular.

METHODS TO STUDY PLANT MATERIAL

The following is the procedure for a scientific study in the laboratory.

Morphological Study

The specimen given for study should be studied for external details either with or without a magnifying glass. Whatever details can be seen are to be recorded. If the specimen requires microscopic study, it should be suitably prepared for the purpose of microscopic observation. There are two types of microscopic preparation.

1. WHOLE MOUNT TECHNIQUE

This is suitable for the purpose of external study of algal material. Take a little of the material, stain it with a suitable stain. Wash the excess stain in water/alcohol. Mount in a suitable medium (temporary or permanent) and observe under the microscope.

2. SECTION CUTTING

For the purpose of anatomical study, sections are to be cut of plant parts. In algae, globule of *Chara*, stolons of *Caulerpa*, receptacles of *Sargassum,* etc., can be cut. Similarly fruiting bodies of fungi and thalli of bryophytes also may be cut in a suitable plane. In higher plants, cylindrical plant parts like root, stem, petiole, etc., may be cut for the study of internal structure.

❮ **3** ❯

MICROTOMY AND ELECTRON MICROSCOPY

INTRODUCTION

Microtomy is a method for the preparation of thin sections for materials and microtomes are scientific instruments meant for taking thin, uniform and entire sections of plant materials like stem, root, leaf, flower, fruit and seed. It is very difficult to take these sections with our hands, whether cross section or transverse and also it takes more time but the chances are lesser and very rare. Thus with help of microtomes, we can take a number of sections of plant materials within a short time. Therefore, the microtomes have been invented as special microtechnique instruments.

PLANES OF CUTTING

In order to study the internal structure from various angles, sections are cut in various planes. The *transverse* section is cut parallel to the long axis (right angles to the transverse axis). The *longitudinal* section may be cut along a radius (Radial longitudinal section-RLS) or along a tangent (Tangential longitudinal section -US).

Cylindrical organs like roots and stems are cut transversely or longitudinally, while dorsiventral organs like thalli of bryophytes, leaves, etc., are usually cut in vertical transverse plane.

TYPES OF SECTION CUTTING

There are usually 2 types of section cutting methods employed for plant parts. These are (a) Free hand sectioning and (b) Microtome sectioning

(a) Free hand Sectioning

For the purpose of laboratory study where slides are not made permanent, usually free hand sectioning method is employed. Free hand sections are cut mostly with the help of a good sharp razor (sometimes safety razor blades also may be used).

Method of Cutting   Open out the razor and bend the handle backwards. Hold the razor between the index finger and thumb. The handle should be free. The remaining fingers of the hand will be touching the back edge of the razor blade.

In order to cut a transverse section of a cylindrical part like stem or root, follow the procedure given below.

The stem or root should be trimmed into a 2-3" long piece. It should be held in the hand in an erect position between the thumb and index finger. The material held in the hand should be exactly at right angles to the razor's edge. The thumb holding the specimen should be slightly at a lower level. Flood the edge of the razor with water. Now using the index finger as the platform move the razor quickly several strokes across the material. During this process sections are cut and they will be collected on the blade of the razor. The sections now may be carefully transferred with the help of a fine brush on to a Petri dish containing water.

For cutting sections of delicate and fragile materials like thalli or leaves a suitable supporting medium has to be used. For this purpose a pith, carrot or potato may be used. These are to be cut into rectangular cubes and split in the center. The material for sectioning should be inserted in the split and sections are cut.

After section cutting, the razor should be wiped dry. It should be greased and encased.

(b) Microtome Sectioning

This is generally employed very frequently in the study of bryophytes, pteridophytes and other higher plants. The microtome sectioning is a long and complicated process. It takes anywhere between 15-20 days to process a material and to obtain a section.

Microtome sectioning involves the following steps with explanation provided for each step:

1. Killing and preserving the material
2. Dehydration and infiltration
3. Paraffin embedding
4. Section Cutting

Killing and Preserving the Material　The material to be sectioned like the stem, leaf, flower, etc., is to be killed and preserved in a suitable fluid.

(a)　Cut the material into bits of suitable size.

(b)　Place the material in small vials containing a suitable fluid.

(c)　Label the vial giving the details of place of collection, name of the material, preservative used and date of collection.

The material should be in the killing fluid at least for six hours (preferably a day).

Dehydration　In order to dehydrate the material, remove it from the preservative, wash thoroughly in water and transfer it to a suitable tube or vial.

Dehydrate the material in alcohol series (Ethyl alcohol) as follows:

Stage 1	50% Alcohol	First day
Stage 2	70% Alcohol	Second day
Stage 3	90% Alcohol	Third day
Stage 4	100% Alcohol	Fourth day
Stage 5	Absolute alcohol and xylol (50/50)	Fifth day
Stage 6	Absolute alcohol and xylol (25/75)	Sixth day
Stage 7	Xylol	Seventh day

When the material is transferred to pure xylol, there should be no fogging (i.e., the solution should not turn milky). If there is fogging repeat the procedure starting from 4th stage.

The material in xylol will be completely dehydraed and looks transparent. It is now ready for embedding by paraffin wax. The purpose of infiltration is to see that the wax should enter into the material completely so that it can easily be cut.

Infiltration   Add a few pieces of paraffin wax everyday for about 4-5 days and then transfer the vial into a hot air oven, while the temperature is adjusted to the melting point of wax. Everyday add a few pieces of wax. Adding of wax should be continued until all the xylol in the vial is replaced by paraffin wax. When the tube is smelt, there should be no smell of xylol. The material is now ready for embedding.

Paraffin Embedding   In order to cut the material, it has to be embedded into a paraffin block. Prepare a paper boat or take a porcelain rectangular dish and smear glycerine to the inner surfaces. Pour molten paraffin into the porcelain dish and empty the contents of the vial into the dish. Before the paraffin solidifies arrange the material in proper rows and in such a manner as desired. Leave it for cooling for 6-8 hours. For quick cooling the dish may be transferred to a bowl containing ice water. Carefully remove the solidified paraffin block from the dish and cut it into suitable pieces. The material is now ready for sectioning.

Sectioning   Sectioning is carried out with the help of an instrument called microtome. There are various types of microtomes like rotary, sliding, freezing, etc. The one that is normally used is a rotary microtome.

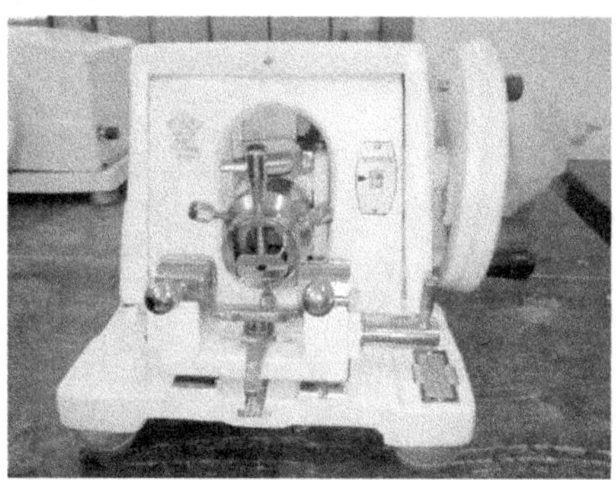

Rotary Microtome - Inner view

TYPES OF MICROTOMES

Some of the commonly useful microtomes are:

1.   Hand microtome

2.   Wood or Jung microtome

3.   Rotary microtome

4.   Rocking microtome, and

5.   Ultramicrotome.

ROCKING MICROTOME

In this microtome, three important parts are recognized. They are:

1.   Razor holder,

2.   Body cylinder with material holder and a pulley system connecting the rocking handle, and

3.   Micron scale operator.

The razor holder is on one side of the microtome and a razor is fixed in it with the help of screw adjustment. Just against the razor is the wax block with material embedded inside this and this is fixed on block holder. The body cylinder is a connecting portion with material holder on one side and the micron adjustment, rocking handle and pulley system at the other end.

The rocking handle is the operation point. When the handle is drawn to the rear end, the block holder comes down and dashes the block on the razor and a section is taken. When the handle is pushed to the distal end, the block holder rises up and brings the block to the original position. When these rocking movements of the handle are made to and fro, a series of sections are taken and these sections unite by end to end and form a wax ribbon. A particular thickness of the section is achieved by fixing the micron scale to a particular point like 10 micron or 15 micron or 20 micron, as the case may be.

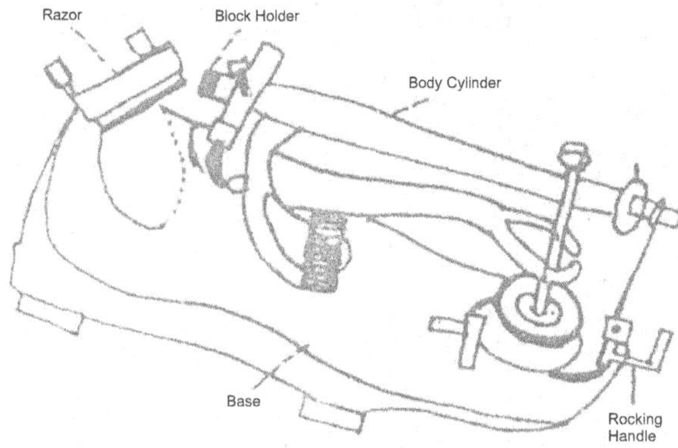

Rocking Microtome

ROTARY MICROTOME

This is more complicated in structure but easier to handle when compared to rocking microtome. This is almost similar in principle and working pattern with rocking microtome. The main difference is the operation point. The operation point or system is a rotating wheel with a handle in the case of rotary microtome and it is a rocking rod with a handle in the case of rocking microtome. In all the other cases of various parts and the working principles, both microtomes are similar. As in rocking microtome, there is razor holder, a block holder, a main body with a rotating wheel and a micron scale adjustment. In the rocking microtome, the main body is kept open, whereas it is completely closed with a metallic cover in the case of rotary microtome.

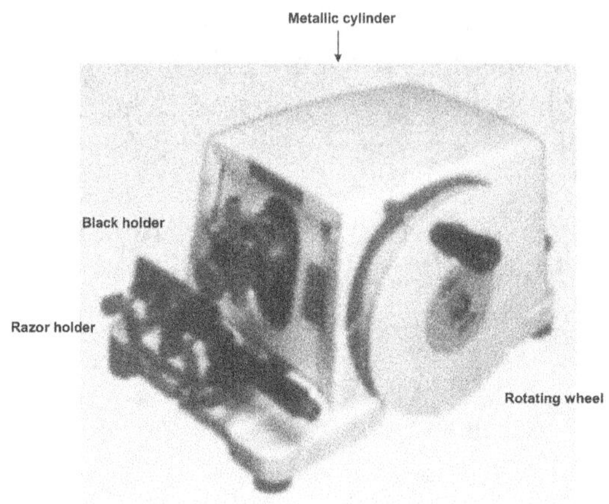

Metallic cylinder

ULTRAMICROTOME

The fast-to-learn, user friendly ultramicrotome allows for most sectioning techniques with precision and speed that yields results of the very highest quality. This motorized ultramicrotome features microprocessor control and retraction. Designed with the operator in mind, the ergonomic configuration actually helps to increase precision by reducing operator fatigue. This microtome looks like completed with microscope orientation head.

MICROTOME SECTIONING

The material to be sectioned by microtome undergoes the following steps:

1. The paraffin block is to be cut into small pieces each having the embedded material.

2. The pieces of paraffin containing the material is to be fixed to wooden 'riders'. For this purpose, heat the top of the rider and the base of the paraffin with a hot knife and fix. Allow it to cool. The paraffin piece is now firmly fixed to the rider. In order to make the edges of the paraffin even, cut the edges and make them parallel.

3. Fix the wooden rider into the microtome so that the paraffin piece with the material is exactly parallel to the edge of the microtome knife.

4. Adjust the knob of the microtome to give the required thickness to the sections.

5. Rotate the wheel of the microtome in a clockwise direction. The cylinder containing the rider moves up and down on the knife edge. Sections are cut in a sequence and get attached to each other forming a paraffin ribbon.

6. Keep a needle in your left hand to hold up the ribbon as it lengthens. When the ribbon is sufficiently long, cut it and transfer it on to a black cardboard. Cut the ribbons into suitable length and transfer to the slide.

7. Smear the slides with an adhesive like egg albumen or gelatin and rub it well.

8. Pour some water onto the slide with the help of dropping bottle.

9.   Transfer the paraffin ribbons on to the slide and drain excess water. Retain a little water on the side.

10.   Place the slide on a slide warmer. The ribbons expand, and so also the sections of the material. Now draw out the water completely and allow the slides to cool.

11.   The slides should be allowed to stand for at least 24 hours where upon the sections will stick to the slide permanently. The slides are now ready for staining.

SECTION STAINING

STAINING PROCEDURE

The plant material (whole mount or sections) are to be stained and mounted properly before they are ready for microscopic observations.

Algal filaments, fungal mycelia, pollen grains and spores, and prothalli of pteridophytes are usually stained with a single stain, safranin or green. In some case aniline blue is used to stain the filaments of algae and mycelia of fungi. The staining procedure is as follows.

Place the material either in a watch glass or cavity slide and add a few drops of stain. The period of staining varies with different materials. But usually a 5-10 minute period is sufficient. Drain out the stain and observe the material under the microscope. If the material has taken less stain, it may be kept in stain for some more period, or if it has taken excess stain it can be removed by washing either with water or 50% ethyl alcohol.

After the material is suitably stained pass it through a series of alcohol grades to dehydrate it. Usually different grades of alcohol (70%, 90%, 100%) taken in a filler are put on the material. After passing through absolute alcohol the material is kept in xylol and it is now ready for mounting.

DOUBLE STAINING PROCEDURE

In the case of most of the algae and fungi where there is little or no tissue differentiation and hence single staining is sufficient. In some cases of plant with relatively more tissue differentiation, single stain is not sufficient to mark out the various tissues clearly. In such cases 2-3 stain combinations are used to stain the tissues. For example lignified and non-lignified tissues are stained with a combination of acidic and basic dyes. This type of staining is also called differential staining where different tissues take up different stains.

The following are some of the popular stain combinations.

Main stain	Counter stain
Haematoxylin	Safranin
	Erythrosine
	Eosin
	Orange
Safranin	Fast green
	Light green
	Aniline blue
	Crystal violet
	Erythrosine

STAINING PROCEDURE FOR FREE HAND SECTIONS

(a) Procedure for Temporary Preparations

Free hand section of root, stem, etc., should be washed thoroughly in distilled water in a watch glass. After repeated washing, the sections should be transferred to watch glass containing the main stain. The sections should be kept in the stain for a period of 30-40 min. After this period, the sections should be transferred to a watch glass containing distilled water. The sections should be washed for 2-3 times to remove the excess stain and observed under microscope. If the main stain is safranin only lignified tissues should retain the stain, others should be destained. If the main stain is specific to non-lignified tissue, only they should retain the stain. Then the sections should be washed thoroughly in water and repeatedly observed under the microscope. The sections can be mildly stroked with a fine hair brush, where excess of stain comes out with distilled water. The sections are now ready for counter-staining. They should be transferred to a watch glass containing the counter stain. Counter stain acts quickly on the tissues. Hence one to two minutes of staining should be sufficient. The sections are now observed under microscope with 15-20% glycerine. After the final staining the tissues would be demarcated clearly with no mixing of the stains. The sections are now ready for mounting.

(b) Procedure for Permanent Preparations

If the sections of plant materials are to be mounted and made into permanent slides, the procedure should be slightly altered so as to include dehydration and clearing with suitable agents. The procedure is as follows:

1.   Stain with the principle stain (safranin, haemotoxylin, etc).

2.   Allow the sections to remain in stain for about 15-20 minutes.

3.   Transfer the sections to water and wash thoroughly. The sections should be stroked mildly until no more stain comes out, and the water should remain clear. If the main stain like safranin is prepared in 50% alcohol then desatining (washing) should be done in 50% alcohol.

4.   The sections should be now dehydrated in alcohol grades of 50%, 70% and 100%.

5.   In each grade of alcohol keep the stain for about 5 minutes.

6.   Transfer the sections from one grade of alcohol to the other with the help of a fine brush. Never use needles as they injure the sections.

7.   Transfer the sections to the counter stain (fast green, etc). The counter stains are usually prepared in 90% alcohol.

8.   Keep the sections in counter stain for about a minute. Wash the sections in 90% alcohol and transfer them to 100% alcohol.

9.   In some case counter stains are prepared in clove oil. In such cases counter staining should be done after the sections are completely dehydrated with absolute alcohol.

10. Sections are transferred completely dehydrated to a solution of 50:50 absolute alcohol and xylol followed by 25:75 absolute alcohol and xylol and ultimately the sections are transferred to pure xylol. Now xylol should not turn milky or turbid. If it turns milky, dehydrate the

sections again through combination of absolute alcohol: xylol series. The sections are now ready for mounting.

STAINING PROCEDURE FOR MICROTOME SECTIONS

The slides with the paraffin ribbons adhering to them have sections of plant material. The processing and staining is as follows:

1. Immerse the slides in a coupling jar containing xylol. Leave them for 10 minutes. The paraffin dissolves in xylol and only the sections remain on the slides.

2. The slides are now kept in 50:50 absolute alcohol and xylol.

3. Pass the slides through a down grade series of alcohol (100%, 90% and 70%).

4. Transfer the slides to pure water and wash them thoroughly to remove traces of alcohol.

5. Keep the slides in a mordant like 4% iron alum. This is necessary only when haematoxylin is used. After about 10 minutes wash the slides in water.

6. Transfer the slides to the main stain haematoxylin, saffranin, etc., and keep them for 15-30 minutes.

7. Remove the slides from stain and wash in water.

8. Observe the slides under microscope. Usually the sections are over-stained.

9. If the main stain used is haematoxylin, destaining can be done with 2% iron alum or with water containing few drops of concentrated HCl. Repeatedly observe the slides under the microscope. Stop de-staining when only the nuclei have retained the stain.

    Caution   When destaining is done with HCl, not more than 5-6 drops should be used for 75 ml of water. Excess acid will eat away the adhesive and the sections come out of the slide.

10. Wash the slides in water overnight if the destaining agent is iron alum, otherwise washing for 30 minutes is enough.

11. Now pass the slides through an upgrade series of alcohol (5 minutes in each grade) 70%, 90% and 100%.

12. When the sections are completely dehydrated stain them with the counter stain (fast green, etc.) for a minute or two. Wash the excess stain in clove oil and transfer the slides to pure xylol. The slides are now ready for mounting.

MOUNTING

MOUNTING MEDIA

Mounting media are of two categories: Temporary mounting media and permanent mounting media.

A temporary mounting medium may be just water or water mixed with glycerine. If the temporary sections are to be retained for 2-3 hours, dilute glycerine (15-20%) is recommended.

Permanent mounting media are used when the slides are to be retained indefinitely. The following are some of the permanent media.

1. ***Canada Balsam*** This is the most widely used mounting medium in botanical preparations. It is available in ready-to-use bottles. It is a gummy substance. If the solution is very thick add little xylol.

    Canada balsam is a resinous product of a gymnosperm, *Abies balsamea*. The gum has very good refractive index and as such it is highly recommended.

2. ***Lactophenol*** Lactophenol can be prepared in the laboratory. Mix equal quantities of phenol crystals, lactic acid, glycerine and distilled water.

3. ***Glycerine Jelly*** B: This is prepared as follows:

Gelatine	1 part
Water	6 parts
Glycerine	7 parts

    Glycerine jelly is suitable even for mounting fresh specimens (even without dehydration). Add gelatine to water, stir thoroughly and warm. Add glycerine and cool the solution. After sometime the solution settles into a jelly.

MOUNTING METHOD

The sections (free hand) are to be mounted on to clean slide in a specific mounting medium. If the sections are microtome preparations they would be already adhering to the slide. In any case the sections are to be covered with a thin coverslip or cover glasses which are readily available in the market. They may be round, square or rectangular. Select the one that is suitable for your preparation.

The sections or objects should be mounted on the centre of the slide. In order to ensure placing of the object on the centre of the slide, the following procedure may be adopted. On a white paper draw a rectangle of the same size as the slide. Join the opposite corners by drawing diagonal lines. The centre of the slide lies where the two lines cross. Place the slide on the rectangle and material or objects may be kept on the central point.

Place a drop or two of the mounting medium. Hold the coverslip at the edges and allow it to touch the edge of the mounting medium at an angle of approximately 45°. Slowly lower the cover glass on the mounting medium with the help of a needle. Take care to see that no air bubbles enter the mounting medium. If there are air bubbles, lift the cover glass and mount again.

Excess mounting medium may be carefully wiped with the help of a clean tissue. Allow the slides to dry. If necessary the slides may be transferred to a hot air oven for drying. The slides should be labelled at one end as shown in the figure. They are now ready for microscopic observation.

RINGING OF SLIDES

Temporary preparations may be kept for few days if the borders are sealed. This procedure is known as ringing. Ringing is done by applying wax, Canada balsam or even nail polish to the edges. Ringing may be done manually or by using a ringing table having a rotary metallic disc.

ELECTRON MICROSCOPY

Basically electron microscopes are of two types.

(i)   Scanning Electron microscope (SEM)

(ii) Transmission Electron microscope (TEM)

1. **SEM** This type of microscope is used to observe only basic architectural view, i.e., morphological–topographical view. For example, phytoplankton, zooplankton, pollen grains.

2. **TEM** This type of microscope is used to observe cytoplasmic inclusion and sub-cellular organelle levels. For example cell wall, chloroplast, starch grains, mitochondria, endoplasmic reticulum, golgi bodies, nucleus, nucleolus, ribosomes, etc.

SCANNING ELECTRON MICROSCOPE (SEM)

The electron gun and lens system of SEM is like TEM but its operation is different.

The specimen to be examined is fixed and dried (by the technique of critical point drying; Karp, 1996) and then coated with a layer of carbon and heavy metal such as gold or gold-palladium—a process called shadowing. This step makes specimen suitable as a target for an electron beam. SEM is used to examine the surface of specimen, *i.e.,* outer cell surface and various processes, extensions and extra-cellular materials. SEM provides a three-dimensional image of a specimen.

In SEM, an extremely fine beam of electrons (5 to 20 nm in diameter) is made at 3-30 kV for scanning a selected area of specimen. In SEM, the electron beam does not pass through the specimen. The condenser lens focuses a fine electron beam on the surface of the specimen. The beam is moved rapidly back and forth by beam deflectors to scan the specimen surface. As the electron beam hits the surface of specimen it excites the specimen molecules to high energy levels. Due to this, secondary electrons are emitted from the metallic surface. These secondary electrons are collected by the positively charged grid. The collectors gives rise to a flash of light in a solid scintillator. The light output is amplified in a photo-multiplier or video amplifier. The signal from the grid is transferred to a television tube which scans and forms the image on the screen. Thus image formation in the SEM is indirect compared to that in the TEM.

The amount of secondary electrons produced depends on the angle of specimen points with the scanning beam. Surface perpendicular to the beam produces maximum electrons while surfaces at greater angles release less electrons. Thus, the number of electrons produced depends on the three-dimensional shape of the specimen surface (or surface topology, i.e., the crevices, hills and pits of the specimen). Accordingly the image contains bright (which correspond the elevations or ridges in specimen surface) and darker regions (which correspond to the valleys). These shadows give the image a three-dimensional appearance.

The resolving power of SEM is comparatively less than that of transmission electron microscope. It has an effective magnification of up to 20,000 times.

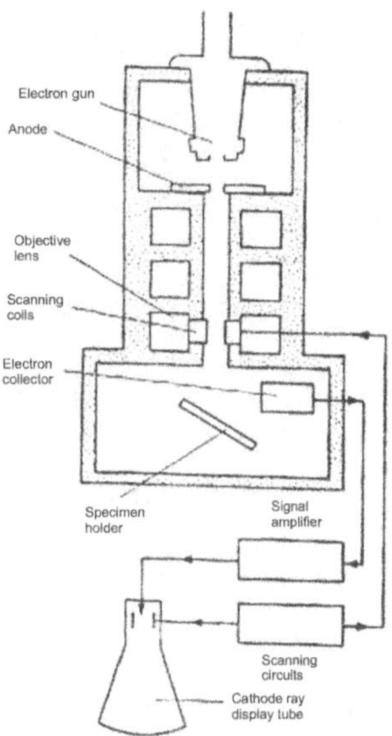

Diagrammatic representation of SEM view

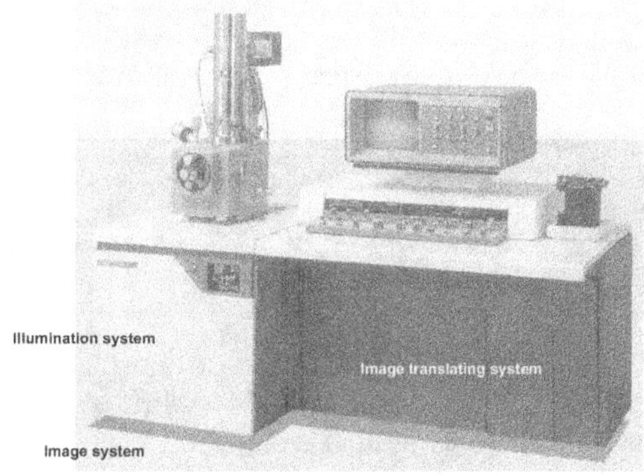

Scanning Electron Microscope

PREPARATION OF MICROALGAE FOR SCANNING ELECTRON MICROSCOPIC STUDIES

Collection of microalgae samples with mud from the field

↓

Wash the sample repeatedly with distilled water

↓

Make the sample almost dry by dehydration

↓

Add equal amount of concentrated HCl to the sample volume
for removal of calcium salts.

↓

Add four times of concentrated H_2SO_4 to the sample volume for
the removal of organic matter.

↓

Heat the sample gently and cool.

↓

Wash the sample repeatedly with distilled water and remove
as much of water as possible.

↓

Reagents included glutaraldehyde, phosphate buffer (pH 6.8), and alcohol. For SEM
study the microalgae were fixed in primary fixative 3% glutaraldehyde.

↓

The fixed samples were given 3 washes thoroughly in 0.1 M phosphate buffer (pH 6.8).

↓

They were dehydrated through a graded series of alcohol at 10-15 minutes interval
at 4°C up to 70%.

↓

Then 90% and 100% alcohol were kept in room temperature at 1 h intervals.

↓

Then dehydrated samples treated with critical point drier (CPD) were on a stub and
the specimens were examined with Joel JSM-56010 with INSA-EDS and electron
micrograph was taken selectively from the computer screen.

PREPARATION OF MACROALGAE FOR SEM STUDIES

Collection of macroalgae samples

↓

Fresh material was collected from the sea, washed well with filtered seawater twice
to remove the unwanted impurities, adhering sand particles and epiphytes thoroughly

↓

Wash the sample repeatedly with distilled water

↓

Cut the specimen 1-2 mm length

↓

Reagents included glutaraldehyde, phosphate buffer (pH 6.8), and alcohol. For SEM
study the macroalgae were fixed in primary fixative 3% glutaraldehyde.

↓

The fixed samples were given 3 washes thoroughly in 0.1 M phosphate buffer (pH 6.8).

↓

They were dehydrated through a graded series of alcohol, 15-20 minutes interval at
4°C up to 70%.

↓

Then 90% and 100% alcohol were kept in room temperature at 2 or 3 h intervals.

↓

Then dehydrated samples treated with critical point drier (CPD) were on a stub and
the specimens were examined with Joel JSM-56010 with INSA-EDS and electron
micrograph was taken selectively from the computer screen.

SECTIONING BY MICROTOME

1. Hand microtome

2. Wood (or) jung microtome

3. Wax microtome

    Rotary microtome—completely closed with metallic cover

    Rocking microtome—main body is kept open.

    1. Razor holder

    2. Body cylinder with material holder are a pulley system connecting the rocking handle.

    3. Micron scale operator.

4. Ultra microtome—automatic mode.

SECTIONING

1. Free hand sections	10 µm thickness cut with razor
2. Serial sections	The paraffin method 5 µm (7.5 × 2.5 cm), cleaned with soap water, rinsed with 95% alcohol and dried for evaporation. At least 25-30 mm left at one end for labelling.
3. Semi thin section	2 –5 µm
4. Ultrathin sections	Gray (20-40 nm) Silver (40-60 nm) Gold (60-80 nm) Yellow (80-100 nm) Orange (100-120 nm) Purple (120-140 nm) Green (140-160 nm)

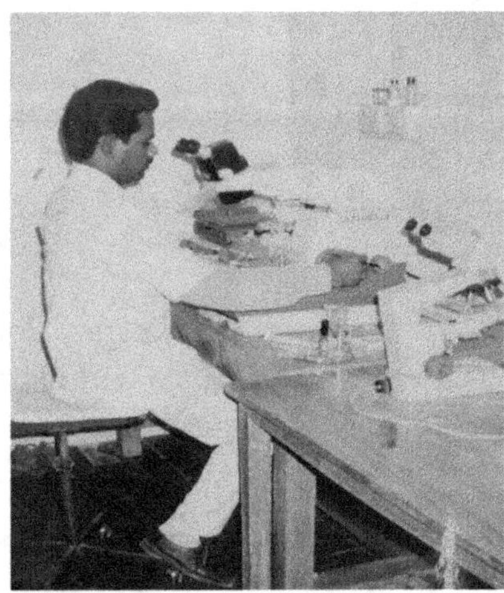

ULTRAMICROTOMY

Trimming of Blocks

Blocks can be trimmed to produce a pyramid form at the tip of the block where the tissue is located. This can be done either manually or mechanically.

Manually it can be done by fixing the block into the holder. Mount it on to the microtome and with a clean razor blade trim the four sides and the tip of the block.

Trimming can also be done with a Pyramitome or block trimmer (Reichert TM 60) fitted with rotating milling cutter.

While trimming the following points must be taken care of.

- Trim as small as possible.

- Do not trim sides of the pyramid too steeply as this reduces the stability.

- If possible trim away all free embedding medium.

- Pyramid should be flat and free from plastic debris.

- Trim the cutting surface of the block in the form of trapezium (△) in order to get serial sections easily.

The trimmed block is fitted in the block holder of the ultramicrotome.

Making Glass Knives

For sectioning diamond or glass knives are used. Diamond knives produce very good sections but they are expensive, therefore most laboratories use glass knives which are less expensive and easily available.

For making glass knives usually Belgium glass strip is used. As the other types of glass have different properties and hardness which tend to break in different manners. Glass knives can be made by glass knife maker by cutting the Belgium glass strip into squares and then cutting the squares diagonally to get triangles. The glass strips are cleaned with acetone before breaking to prevent any traces of grease. Knives with concave edge and irregular or partly broken edges are discarded. Glass knives should be used within days of their having been made. During the period they must be stored in a dust-free container with their cutting edge upwards.

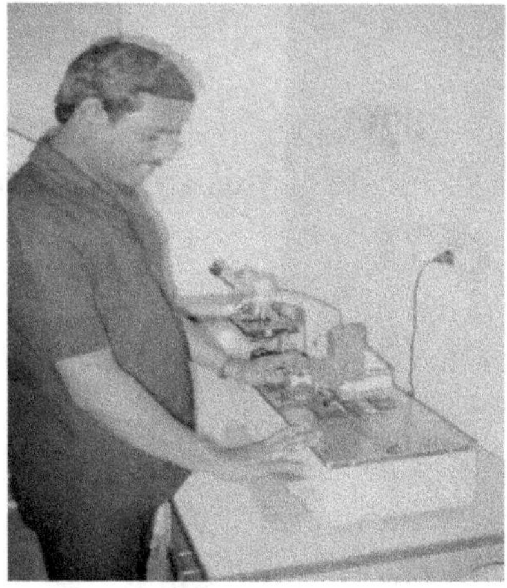

LKB-Knife Cutter

Filling the Boat

To allow the sections to be cut without sticking to the knife edge, a trough (boat) is placed behind the edge. This is filled with water for collecting sections. The boat is either made from tape or from thin metal. It should end at the same height as the knife edge. To make the boat water-tight, seal the lower part of the boat with the knife, using either wax or nail polish.

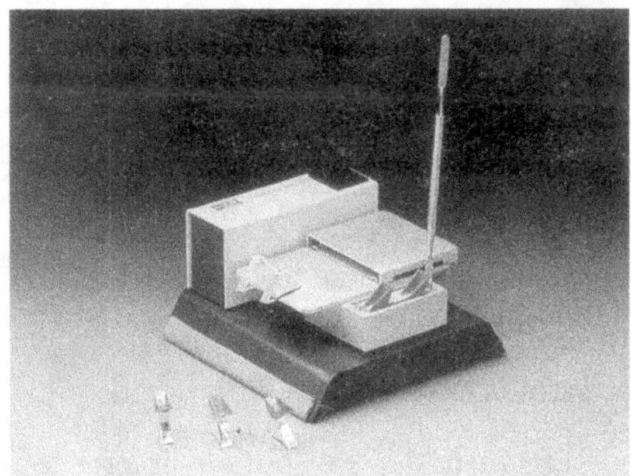

Workstation

A three temperature work station, for convenient slide drying and staining, plus a wax bath for use in mounting LKB Trufs on glass knives is shown in the picture.

Preparation of Semi-Thin (Thick) Sections for Optical Microscopy (Survey Sections)

Before one proceeds with ultrathin sectioning, thick sections (0.5 to 2.0 μm) are cut scanning the tissue under an optical microscope. This enables the electron microscopist to determine the following:

- The state of the embedded specimen.
- The fixation.
- Selection of the area of ultrathin sectioning.
- The size and position of the cutting face for final trimming.

The semi-thin sections are floated in water or 10% acetone, are lifted with a glass rod or a thin brush and placed on a clean glass slide. The slide is placed on a hot plate at about 80°C and dried. Methylene blue/Azure II and/or Toluidine blue are some of the stains commonly used for staining.

METHYLENE BLUE/AZURE II

Prepared by mixing equal parts of 1% methylene blue in 1% Borax and 1% Azure II.

Toluidine Blue

1% solution in 1% Borax

Stain sections for 0.5 to 2.0 mm (depending on the thickness). Wash thoroughly in running water and then in distilled water. Dry and mount cover-slip with DPX.

Observe under the light microscope.

Preparation of Ultrathin Sections

After scanning the section under the optical microscope the area to be examined under the TEM is selected and the blocks are further trimmed (hand trimming is advised).

Use fresh glass knife for cutting ultrathin sections. On the liquid in the trough the ultrathin sections show interference colours. This makes it possible to determine the thickness of the sections.

Gray: 60 nm (600 Å) optimal for high resolution work

Silver: 60–90 nm ideal for most purpose

Gold: 90–150 nm useful for low magnification and autoradiography

Purple blue, green and yellow range from 150-320 nm; very thick sections and are suitable for scanning and transmission electron microscopy.

[1 mm $=$ 1000 μm, 1 μm $=$ 1000 nm; 1 nm $=$ 10Å (1Å $=$ 10^{-7} mm)]

The ultrathin sections floating on the water are stretched by exposing them to chloroform vapours. The sections are lifted from below on specially made metal grids. Grids made of copper are commonly used. Grids made of nickel, stainless steel, gold and platinum are also available which are used for specific purposes.

The grids are available as meshed or holed ones and grids of 100 to 300 mesh size are most commonly used. One of the surfaces of the grids is shiny and the other matted. The sections are lifted using the matted surfaces of the grid which affords sticking of the sections firmly.

If the grids of large mesh size of holed grids are used a coating of film is essential to render support to the sections. Film coating is done on the matted surface of the grid.

3% formvar (Polyvinyl formaldehyde) in ethylene dichloride and/or carbon are generally used for coating the grid. Carbon coating is done by a process of evaporation under vacuum on a grid already coated with formvar. After double coating, the formvar is dissolved either with ethylene dichloride or chloroform leaving behind the carbon film. Coated grids are useful for supporting particulate specimen like viruses and bacteria.

Staining Ultrathin Sections

To obtain a good contrast a double staining method using uranyl acetate and lead citrate is routinely followed for histological studies.

Uranyl Acetate   Prepare a saturated solution by adding excess of uranyl acetate (analar) to 10 ml of filtered 50% ethanol in a 15 ml centrifuge tube. Shake vigorously for 2 min. and spin down hard to allow the excess of uranyl acetate to settle down. The solution is ready for use. Stopper and store at 4°C.

Lead Citrate   Add one half pellet of sodium hydroxide to 12 ml of double distilled water in a 15 ml centrifuge tube. Shake well to dissolve the NaOH. Add 50 mg of lead citrate (analar) and shake well for 2 min. and centrifuge.

The solution is ready for use. Stopper and store at 4°C.

Staining Procedure   Filter or centrifuge the stains before use. Pipette out a small amount of uranyl acetate solution in a clean watch glass. Place the grid carrying the section down onto the stain. Place a wooden or card board cover as the staining is effective when carried out in dark. Allow the stain to remain for 10-15 min. Take each grid and wash in 2 lots of 50% ethanol and 2 lots of double distilled water with continuous agitation. Dry carefully on a filter paper (do not allow the section carrying surface of the grid to come into contact with the filter paper).

Place a few ml of lead citrate in a watch glass and place the grids onto the stain and stain for 5 to 10 min. Wash each grid briefly in 0.02 M sodium hydroxide and then in 2 lots of double distilled water. Dry and store the grids in grid boxes or a well covered Petri dish.

The sections are ready for observation in the transmission electron microscope.

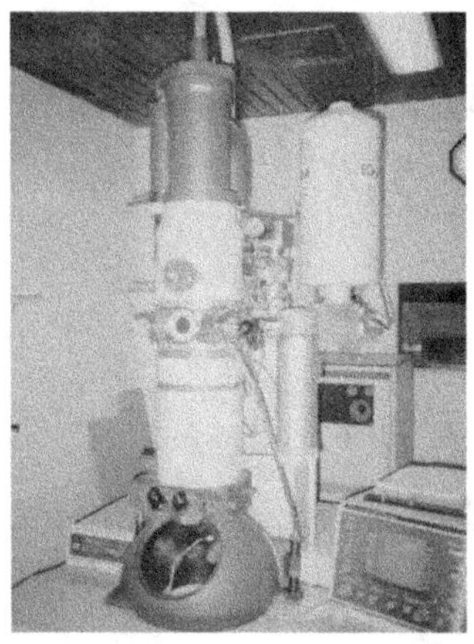

Transmission Electron Microscope CM Philips 12

PREPARATION OF MATERIALS FOR ELECTRON MICROSCOPY (TEM)

Primary fixation	2% glutaraldehyde at 4°C overnight
↓	
Rinse in buffer	Three times each 15 min. interval at 4°C
↓	
Post fixation	1% OsO_4 overnight at 4°C (duration depending
↓	upon the nature of the material)

Dehydration   In Acetone/Alcohol

20%	30 min. interval at 4°C
35%	30 min. interval at 4°C
50%	30 min. interval at 4°C
70%	30 min. interval at 4°C
90%	1 h interval at room temperature (RT)
100%	2 h interval at RT

Infiltration

Acetone:Araldite

75%   3:1

50%   1:1

25%   1:3

Embedding   On silicon rubber mould (or) beam capsule

↓

Polymerization

↓

Semi-thin section   2-5 μm thickness

Ultrathin section 60-90 mm (silver and gold colour)

↓

Staining   Positive staining

Uranyl acetate + lead citrate

Observation under TEM

RECORD OF PRACTICAL WORK

Practical work carried out in the laboratory should be recorded. The botany practical record book has two types of sheets, viz., interleaves and drawing sheets. Interleaves are meant to write the notes pertaining to the diagrams. The drawing sheets which are thicker than the interleaves are used for diagrams. The following points may be useful in maintaining a good record.

1.   Use good pencils (H, HB) with sharp points and a good quality eraser.

2.   Leave a minimum of one inch margin on all the sides.

3.   Always draw diagrams of only one plant in a page. Do not mix up diagrams of other plants.

4.   On the left hand corner of the drawing sheet write the date of the experiment.

5.   Give the complete systematic position of the member on the top of the drawing sheet.

6. For microscopic organisms draw first the habit sketch followed by low power and high power view/s. High power view should show the cellular details.

7. In the case of ariatomic21 drawings, draw the ground plant first, followed by a sector enlargement to show the cellular details.

8. All parts of the diagrams must be properly labelled.

9. Stages of reproduction, if any, must be drawn after the details of vegetative structure.

10. The description corresponding to the diagrams must be written on the opposite interleaf.

FIELD COLLECTION OF PLANTS

Specimens of algae, fungi and bryophytes can be collected in the field for the purpose of laboratory study. Field trip is an integral part of botanical study and all students must participate in it. It is in the field study that the students will come to understand the natural growth, habitat and environmental association of plants. During a field trip to collect specimens, the following items are to be carried:

1. Pocket lens

2. Fixative - FAA or Carnoy's fluid

3. Collection bottles

4. Blotting paper

5. Polythene bags

6. Field note book

7. Labels.

Collection of seaweeds from their natural habitat

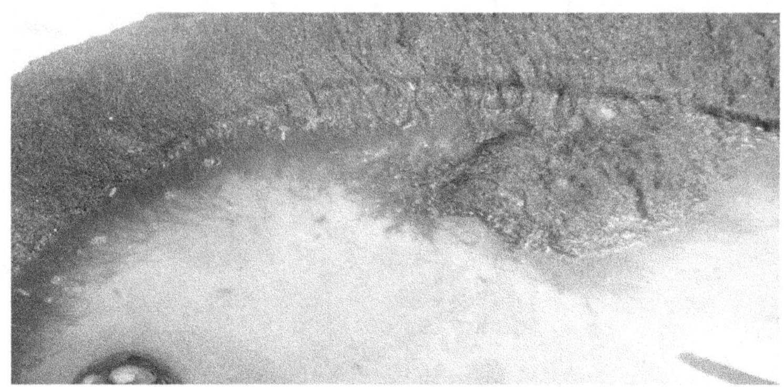

Chlorophyceae – *Enteromorpha* sp.

Phaeophyceae – *Padina* sp.

Rhodophyceae – *Gracilaria* sp.

When the specimens are collected they can be fixed and preserved for the purpose of laboratory study. Use the required preservative and immerse the specimens in it. Label the bottle showing details of:

- Date of collection

- Location

- Preservative fluid (name) used

- Name of the plant

As it is necessary for the purpose of identification, details of habitat, associations with other plants if any, colour of the fresh specimen, etc., should be recorded in the field note book.

Turbinaria sp. (Phaeophyceae) – thallus attached to the rocky substratum

$$\approx \quad 4 \quad \approx$$

HERBARIUM

Herbarium is the collection of pressed, dried plant specimen mounted on a specified sheet, identified and arranged by an approved system of classification. They are arranged in a sequence of order accepted by the rules and regulations of ICBN (International Code of Botanical Nomenclature).

Materials required for preparing herbarium are as follows:

- Plastic trays
- Forceps
- Specimen mounting paper (herbarium sheets)
- Cheese cloth
- Blotting paper
- Herbarium wooden press
- Painting brush
- Pencils, knife, etc.
- Polyethylene bags

PROCEDURE FOR PREPARING HERBARIUM

- Fresh specimen should be cleaned of sand particles, rocks, shells, mud and other adhering materials and epiphytes.

- A tray containing fresh water (half filled) should be taken and specimen to be mounted be placed in the water.

- A herbarium sheet, size smaller than the tray to be inserted from below the specimen and then spread the specimen on the herbarium sheet with the help of brush in such a way that overlapping of the specimen is minimized.

- After mounting the specimen on the herbarium sheet, sheet is lifted slowly and tilted to one side to allow water to drain gradually without disturbing the mounted specimen.

- Remove the sheet and properly arrange the specimen with the help of forceps or needle if required.

- To blot dry, herbarium sheets are placed on the newspaper sheets or blotting paper to remove the remaining water from the herbarium.

- A cheese cloth is placed on the top of the specimen in such a way that it covers entire specimen.

- Now place another sheet of the blotting paper over the herbarium sheet.

- Once, all the specimen to be mounted are ready, herbaria are piled one above the other and then placed between the two sheets of the wooden press.

- The press is tied tightly with appropriate pressure by a rope.

- The press is kept at room temperature for 24 h. After 24 h, blotting papers are required to be replaced.

- The process of replacing blotting papers is repeated till the time specimen is free of moisture.

- On drying of the specimen, the specimen gets attached to the paper due to the phycocolloid present in the seaweed.

- The cheese cloth is carefully removed and herbarium sheet is properly labelled containing collection number, name of the specimen, locality, date of collection and other ecological details.

- Sheets can be placed in the polyethylene bags and sealed and stored.

IDENTIFICATION OF SEAWEED SPECIES

Although, identification of the seaweed species is time consuming and tedious, it is interesting. Beginners should get familiar, first with herbarium specimen from the museum or reference collection before going for the field collection. Colour and morphological differences between different genera/ species and taxonomic characteristic are required to be carefully studied. Only practice of handling and distinguishing the plants in the natural habitat will help a great deal in learning seaweed identification.

Taxonomic identification key should be followed to identify the seaweed specimen. The taxonomic description of the specimen and anatomical characteristic of the specimen to be identified should be referred from the books. Once you identify the specimen tentatively following the key, you should compare it with the herbarium from reference center. Later, you may once again get it confirmed from the expert in the field. Some of the seaweed species, particularly, filamentous seaweed are difficult to identify. In such cases chemotaxonomy or genetic approach could be employed.

MOUNTING THE SPECIMEN

Many specimens will remain attached to the herbarium sheet following drying due to the presence in the algal walls and intercellular spaces of colloidal "glues". The coarser, non-gelatinous forms (e.g., some Phaeophyta) may not remain attached after drying and may require "glue".

Any good clear-drying glue may be adequate, such as white glue or a white PVA resin. However due to problems with white glue becoming soft/sticky again (under humid conditions), it is preferable to use "tin" paste applied in spots, to the underside of the specimen. Gummed linen herbarium tapes may also be used to "strap" the specimen(s) to the sheet.

APPLYING SPECIMEN LABEL

Using good quality (100% rag acid-free) herbarium label paper, complete the label and affix by means of a clear-drying cement (tin paste), to the lower right-hand corner of the sheet. Avoid gummed labels on poor quality paper.

The completed herbarium sheet should include a label, and may also have museum, as well as annotation notes written directly on the sheet or on separate labels.

HERBARIUM LABELS

There are numerous variations in style; however a herbarium label should be on 100% rag paper and should contain the following information:

- Geographic area of collection (i.e., name of island, country, state)
- Binomial, including author(s)
- Where collected, including latitude, and longitude if possible
- Depth, substratum type, etc., including how collected (Scuba, dredge, submersible, etc.)
- Specific ecological information
- Collector; date of collection
- Collector's field number for specimen or collection
- Person who identified the specimen

It is important to remember that the information gathered along with the specimen is just as important as the specimen itself. A collection notebook should be kept that details the collections' locality, habitat, water conditions, etc. If necessary, it is possible to refer back to information in this notebook.

Washing the seaweed in water to remove salt and adhering sand particles

Mounting the seaweed on the Herbarium sheet and draining of excess water from the sample

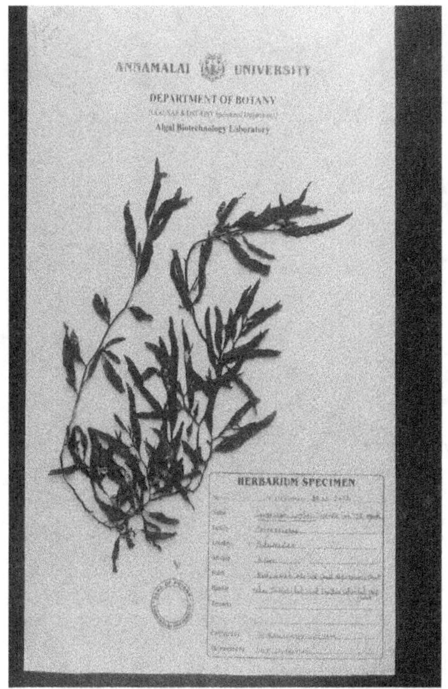

The seaweed mounted on the herbarium sheet, pressed, dried, labelled and deposited in the Department of Herbarium

$$\text{\ae} \quad 5 \quad \text{\ae}$$

CLASSIFICATION OF ALGAE

Algae exhibit great variation in their thallus structure and organization with respect to their size, shape, number and structure of cell, pigmentation, nutrition, reserve food, growth, movement and reproduction from species to species.

CLASSIFICATION OF ALGAE BASED ON THEIR SIZE

Algae that are less than 2 micron (μm) in diameter and can be viewed only with the aid of microscope are known as microalgae and those which are large enough to be seen with the unaided eye are referred to as macroalgae. Microalgae include both unicellular and multicellular forms.

Unicellular Algae

The microalgal species which occur as solitary cells are referred to as unicellular algae, species like chlorella occur as non-motile solitary cells which are described as the coccoid habit by Fritsch (1935) and as proto-coccoidal type by Round (1973) whereas species like Chlamydomonas are motile. Motile cells may possess one or more flagella or they move by gliding. A wide variety of forms exists among unicells, including those contained within a gelatinous sheath, with intricate cell walls, having flexible cell shapes, with two flagella of equal or unequal length. Round (1973) divided the unicellular thalli of algae into three categories, i.e., rhizopodial type, protococcoidal type and flagellated unicells.

Multicellular Algae

Multicellular thalli are of five major types and are described as colonial, filamentous, siphonaceous, foliose and heterotrichous forms. A colony with definite number of cells and having a constant shape and size is said to be coenobium which could be motile like volvox or non-motile like pediastrum. Huge masses of cells that remain embedded irregularly in mucilage constitute palmelloid colony as in the case of microcystis. A number of cells joined end to end by the production of localized mucilage forms a dendroid as in dinobryon. An aggregation of cells in an end-to-end fashion contained in a common gelatinous matrix but are not directly connected to each other constitute a pseudo-filament as that of chroodactylon. A series of cells that are arranged in an end-to-end manner, where adjacent cells share a common cross wall forms a trichome. A trichome surrounded by a sheath is called a filament.

When the cells are arranged in a single series, the filament are said to be uniseriate as in spirogyra and when there are more than one series of cells it is said to be multiseriate like bangia, filaments may be unbranched as that of zygnema or branched like in cladophora, false branches formed by the fragmentation and continued growth of one or both fragments are seen in some cyanobacteria such as scytonema.

A tube-like thallus with a central vacuole and many nuclei in the cytoplasm but without any septa is said to be siphonous and coenocytic as seen in vaucheria. The thallus of algae like Batrachospermum is designated as uniaxial as it has only one main axis and many lateral branches whereas the thallus of algae like polysiphonia consisting of large numbers of similar filaments in close juxtaposition giving the appearance of many axes is said to be multiaxial thalli that superficially resemble parenchyma, but which are actually composed of oppressed filaments or amorphous cell aggregates. If the thallus is many layer thick, it is described as parenchymatous which could be membranous as in Ulva or tubular as in Enteromorpha.

Freshwater algae exhibit all of these morphologies, but the macroscopic parenchymatous and pseudoparenchymatous forms tend to be smaller than those found in marine systems. Algae, on the basis of their habitat can be recognized into four major catagories namely aquatic algae, aerial and subaerial algae, terrestrial algae and algae of unusual habitat.

CLASSIFICATION OF ALGAE BASED ON THEIR AQUATIC HABITAT

Aquatic algae are found growing in aquatic environments including fresh water, brackish water, and seawater.

a. **Fresh water algae** Species of oedogonium, chara and zygnema, etc., grow in the still fresh water bodies whereas cladophora grow in the running fresh water bodies.

b. **Brackish water algae** Anabaena, Oscillatoria, etc., are found in brackish waters.

c. **Marine algae** Algae like Gracilaria, Fucus, Padina, etc., live in the seawater.

PHYTOPLANKTONIC ALGAE

Aquatic organisms that float on the surface of water are called plankton. Those belonging to the plant kingdom are referred to as phytoplankton. Thousands of algal species live as phytoplankton in both fresh water and marine water ecosystems. Picoplankton is the smallest with a size of less than 2 micrometers in diameter. Nanoplankton is an intermediate sized microalgae with size ranging from 2-20 micrometer. Microplankton is the largest phytoplankton and includes those algae greater than 20 micrometer in diameter.

CLASSIFICATION OF ALGAE BASED ON THEIR HABITAT ON LAND

Aerial and Sub-Aerial Algae

They grow in open substrata like the trunks of trees, walls, fencing wires, rocks, etc., that are exposed to open atmosphere to varying degrees. They may be further classified as follows:

Epiphyllophytes Such algae grow epiphytically upon leaves of plants.

Corticolous   Algae like trentipohlia which inhabit the bark of trees.

Epiphloephytes   These algae grow on the barks of trees mixed with many mosses and liverworts, e.g., Phormidium, Scytonema, Haplosiphon and Schizothrix.

Epizoophytes   These algae are found even on the bodies of land animals. Certain chaetophorales are found even on the hairs of sloth.

Lithophytes   Algae growing on the rocks and walls. Species of scytonema, nostoc, etc., grow at the wet walls or rocks in rainy season.

Epilithic Algae   They grow on a wide range of hard surfaces and artificial substrates such as glass and plastic. Epilithic communities in turbulent littoral habitats are distinct from epiphytic communities within the same lake and comprise species known mainly from streams, such as chamaesiphon spp., and Gongrosira incrustans.

Terrestrial Algae

These algae are found growing in the soils. The soil flora is collectively known as edaphophytes (Prescott, 1969). Those algae like Mesotaenium, Botrydium, Protosiphon, etc., growing on the soil surface are described as saphophytes or epiterranean algae become partially heteromorphic in darkness (Parker, 1971). Some algae may grow up to a depth of more than one meter and they need proper moisture, illumination and nutrition. Friedmann and Ocampo (1976) observed small sized algae in desert soils. Some common terrestrial forms are Rivularia, Scytonema, Vaucheria, etc.

Friedmann *et al.* (1976) classified the desert soil algae into 5 categories.

- **Endedaphic**   Algae living in the desert soil.

- **Epidaphic algae**   Algae on the surface of desert soil.

- **Hypolithic**   Algae living on the lower surface of stones on desert soil.

- **Chasmolithic**   Algae living in the rock fissures in desert soil.

- **Endolithic**   Algae, which penetrate rock as they grow.

Episammic Algae or Psammon

These are the algae growing on sandy beaches, e.g., Phormidium, vaucheria and many diatoms. Most conspicuous components of episammic algal flora are the diatoms which predominantly belong to non-motile genera (Round, 1981)

Epipelic Algae

Algae growing on the surface of sediments ranging from fine muds to coarse sands constitute epipelic algae. They mainly belong to pinnate diatoms, blue green desmids and motile members of other groups and are seen in both fresh water and marine habitats.

Algae of Unusual Habitat

Cryophytes   These algae are found on ice and snow. These algal form, cause red snow, green snow, yellow snow, yellowish green snow and violet snow colors to the mountains. Their growth depends on the pH of the ice or snow. The green snow in arctic region is caused by species of Chlamydononas, Ankistrodesmus and Mesotaenium and red snow is caused by species of Scotiella and Gloeocapsa.

SYMBIOTIC ALGAE

These algae are found living in symbiotic association with plants or animals. The algae which spend most of their life cycles attached to the surface of other organisms are described as epibionts.

Endophytic Algae

Algae growing within or among the cells of other plants, e.g., *Nostoc* in the thallus of *Anthoceros, Anabaena cycadearum* in the coralloid roots of the cycas, *Anabaena azollae* in the freshwater fern, *Azola azolla.*

Ahmadjian (1967) gave a detail guide to algae occurring as algal components in lichens. Nostoc, Trebouxia, Coccomyxa, Trentepohlia, Uroccus, Phycopeltis and Prasiola are common phycobionts of lichens.

Epizoic and Endozoic Algae

Algae growing on the body surface of animals are called epizoic and those which grow within animals are known as endozoic. Cladophora on snails and zoochlorella within hydra are examples of symbiosis on animals. *Cladophora crispate* grows epizoically on the shells of molluscs, whereas *Stigeoclonium*, a chlorella-like green algae are seen within paramecium, hydra and molluscs.

PARASITIC ALGAE

Some of the algae occur parasitically on plants and animals. A few of them are listed below:

Algae Parasitic on Plants (Phytoparasitic Algae)

Cephaleuros grows on the leaves of plants such as piper, guava, magnolia and theasinensis. Holcomb (1975) reported 115 species, susceptible to *Cephaleuros virescens*.

a)   Phyllosiphon on the leaves of *Arisarum vulgare*

b)   Rhodochytrium on ragweed leaves

c)   *Polysiphonia fastigiata* occurs semi-parasitically on *Ascophyllum nodosum*.

Algae Parasitic on Animals (Zooparasitic Algae)

The cyanobacterium *Phormidium corallyticum* is a dominant component of the pathogenic microbial consortium known as black band disease which kills coral colonies on reefs world e (Richardson 1997). *Prototheca wickerhamii* and *P. zopfii* are non-photosynthetic members of the green algae that can cause infections in humans and cattle.

CLASSIFICATION OF PLANKTON

Welch (1952, P. 225) lists the various classification of plankton based on quality, size, local environment, distribution, origin, etc.

1.   Phytoplankton: Proper chlorophyll bearing plankton

2. Saproplankton: Bacteria on fungi

3. Zooplankton: Animal plankton

On the Basis of Size

a. Nanoplankton: The larger unit plankton visible to the unaided eye

b. Net plankton (Mesoplankton): Plankton secured by the plankton net equipped with no. 25 silk bolting cloth 9 of mesh size 0.03 to 0.04 mm.

c. Nanoplankton (micro plankton): Very minute plankton not secured by the plankton net with no. 25 silk bolting cloths.

On the Basis of Local Environment Distribution

Limnoplankton   - Lake plankton

Rheoplankton   - (Potamoplankton) Running water plankton

Heleoplankton   - Pond plankton

Haliplankton   - Salt water plankton

Hypalmyroplankton   - Brackish water plankton

On the Basis of Origin

Autogenic plankton   - Plankton produced locally

Allogenic plankton   - Plankton introduced from other localities

On the Basis of Content

Euplankton   - True plankton

Pseudo plankton - Debris mingled in plankton

On the Basis of Life History

A. Holoplankton   Organisms free-floating throughout their life.

B. Meroplankton   Organisms free-floating at certain times in the stages of their life-cycle.

On the Basis of Water Movement

There are two different types of water bodies. They are lentic and lotic. Lentic and lotic water bodies in turn comprise of different water bodies which are given as follows:

Lotic	Lentic
Springs	Ditches
Streams	Pools
Canals	Puddles
Waterfalls	Ponds
River	Reservoirs
Rivulets	Lakes
	Paddy fields

CLASSIFICATION OF ALGAE BASED ON PIGMENT

Beginning in the 1830s, algae were classified into major groups based on colour—e.g., red, brown, and green. The colours are a reflection of different chloroplast pigments, such as chlorophylls, carotenoids, and phycobiliproteins. Many more than three groups of pigments are recognized, and each class of algae shares a common set of pigment types distinct from those of all other groups.

1. Division Chlorophyta (green algae)

Chlorophylls *a* and *b*; starch stored inside chloroplast; mitochondria with flattened cristae; flagella, when present, lack tubular hairs (mastigonemes); unmineralized scales on cells or flagella of flagellates and zoospores; conservatively, between 9,000 and 12,000 species.

Classn Chlorophyceae   Primarily freshwater; includes *Chlamydomonas, Chlorella, Dunaliella, Oedogonium,* and *Volvox.*

Class Charophyceae Includes the macroscopic stonewort *Chara*, filamentous *Spirogyra*, and desmids.

Class Pleurastrophyceae Freshwater and marine; includes marine flagellate *Tetraselmis*.

Class Prasinophyceae (Micromonadophyceae) Paraphyletic, primarily marine; includes *Micromonas* (sometimes placed in Mamiellophyceae), *Ostreococcus*, and *Pyramimonas*.

Class Ulvophyceae Primarily marine; includes *Acetabularia*, *Caulerpa*, *Monostroma*, and sea lettuce (*Ulva*).

2. Division Chromophyta

Most with chlorophyll *a*; one or two with chlorophyllide *c*; carotenoids present; storage product beta-1, 3-linked polysaccharide outside chloroplast; mitochondria with tubular cristae; biflagellate cells and zoospores usually with tubular hairs on one flagellum; mucous organelles common.

Class Bacillariophyceae (diatoms) Silica cell walls, or frustules; centric diatoms commonly planktonic and valves radially symmetrical; pennate diatoms, usually attached or gliding over solid substrates, with valves bilaterally symmetrical; primarily in freshwater, marine, and soil environments; at least 12,000 to 15,000 living species; tens of thousands more species described from fossil diatomite deposits; includes *Cyclotella* and *Thalassiosira* (centrics) and *Bacillaria*, *Navicula* and *Nitzschia* (pennates).

Class Bicosoecaceae May be included in the Chrysophyceae or in the protozoan group Zoomastigophora; colourless flagellate cells in vase-shaped loricas (wall-like coverings); cell attached to lorica using flagellum as a stalk; lorica attaches to plants, algae, animals, or water surface; freshwater and marine; fewer than 50 species described; includes *Bicosoeca* and *Cafeteria*.

Class Chrysophyceae (golden algae) Many unicellular or colonial flagellates; also capsoid, coccoid, amoeboid, filamentous, parenchymatous, or plasmodial; many produce silica cysts (statospores); predominantly freshwater; approximately 1,200 species; includes *Chrysamoeba, Chrysocapsa, Lagynion,* and *Ochromonas.*

Class Dictyochophyceae   Predominantly marine flagellates, including silicoflagellates that form skeletons common in diatomite deposits; fewer than 25 described species.

Order Pedinellales   When pigmented, has 6 chloroplasts in a radial arrangement; flagella bases attached almost directly to nucleus; includes *Apedinella, Actinomonas, Mesopedinella, Parapedinella,* and *Pteridomonas.*

Order Dictyochales (silicoflagellates)   Typically with siliceous skeletons like spiny baskets enclosing the cells; flagella bases attach almost directly to nucleus; silicoflagellate skeletons common in diatomite deposits; includes *Dictyocha, Pedinella,* and *Pseudopedinella.*

Class Eustigmatophyceae   Mostly small, pale green, and spherical; fewer than 15 species; *Eustigmatos* and *Nannochloropsis.*

Class Phaeophyceae (brown algae or brown seaweeds)   Range from microscopic forms to large kelps more than 20 metres long; at least 1,500 species, almost all marine; includes *Ascophyllum, Ectocarpus, Fucus, Laminaria, Macrocystis, Nereocysti-s, Pelagophycus, Pelvetia, Postelsia,* and *Sargassum.*

Class Prymnesiophyceae (Haptophyceae)   Many with haptonema, a hair like appendage between two flagella; no tubular hairs; many with organic scales; some deposit calcium carbonate on scales to form coccoliths; coccolithophorids may play a role in global warming because they can remove large amounts of carbon from the ocean water; predominantly marine and planktonic; approximately 300 species; more fossil coccolithophores known; includes *Chrysochromulina, Emiliania, Phaeocystis,* and *Prymnesium.*

Class Raphidophyceae (Chloromonadophyceae)   Flagellates with mucocysts (mucilage-releasing bodies) occasionally found in freshwater or marine environments; fewer than 50 species; includes *Chattonella, Gonyostomum, Heterosigma, Psammamonas,* and *Vacuolaria.*

Class Synurophyceae   Previously placed in Chrysophyceae; silica-scaled; unicellular or colonial flagellates sometimes alternating with capsoid benthic stage; cells covered with elaborately structured silica scales; approximately 250 species; *Mallomonas* and *Synura*.

Class Xanthophyceae (yellow-green algae) Primarily coccoid, capsoid, or filamentous; mostly in freshwater environments; about 600 species; includes *Botrydium*, *Bumilleriopsis*, *Tribonema*, and *Vaucheria*.

3. Division Cryptophyta

Unicellular flagellates.

Class Cryptophyceae Chlorophyll *a*, chlorophyllide c_2, and phycobiliproteins; starch stored outside of chloroplast; mitochondria with flattened cristae; tubular hairs on one or both flagella; special ejectosomes in a furrow or gullet near base of flagella; cell covered with periplast, often elaborately decorated sheet or scale covering; nucleomorph may represent reduced nucleus of symbiotic organism; approximately 200 described species; includes *Chilomonas*, *Cryptomonas*, *Falcomonas*, *Plagioselmis*, *Rhinomonas*, and *Teleaulax*.

4. Division Rhodophyta (red algae)

Predominantly filamentous; mostly photosynthetic, a few parasitic; photosynthetic species with chlorophyll *a*; chlorophyll *d* present in some species; phycobiliproteins (phycocyanin and phycoerythrin) in discrete structures (phycobilisomes); starch stored outside chloroplast; mitochondria with flattened cristae; flagella completely absent; coralline red algae contribute to coral reefs and coral sands; predominantly marine; approximately 6,000 described species; includes *Bangia*, *Chondrus*, *Corallina*, *Gelidium*, *Gracilaria*, *Kappaphycus*, *Pa-lmaria*, *Polysiphonia*, *Porphyra*, and *Rhodymenia*.

5. Division Dinoflagellata (Pyrrophyta)

Taxonomy is contentious. Predominantly unicellular flagellates; approximately half of the species are heterotrophic rather than photosynthetic; photosynthetic forms with chlorophyll *a*, one or more chlorophyllide *c* types, and peridinin or fucoxanthin; mitochondria with tubular cristae and flagella without tubular hairs; ejectile trichocysts below surface in many members; many with cellulosic plates that form a so-called armour around cell; some bioluminescent, some containing symbionts;

resting (interphase) nucleus contains permanently condensed chromosomes; several produce toxins that either kill fish or accumulate in shellfish and cause sickness or death in humans when ingested; more than 1,500 species described, most in the class Dinophyceae; includes *Alexandrium, Ceratium, Dinophysis, Gonyaulax, Gymnodinium, Nocti-luca, Peridinium,* and *Polykrikos.*

6. Division Euglenophyta

Taxonomy is contentious. Primarily unicellular flagellates; both photosynthetic and heterotrophic.

Class Euglenophyceae   Chlorophylls *a* and *b*; paramylon stored outside chloroplasts; mitochondria with paddle-shaped cristae; flagella lack tubular hairs, but some with hair like scales; pellicle covering of sliding sheets allows cells to change shape; approximately 1,000 described species; includes *Colacium, Euglena, Eutreptiella,* and *Phacus.*

❧ **6** ☙

DESCRIPTION OF ALGAL SPECIES

CHLOROPHYTA

CHLOROPHYCEAE

General Characteristics Features of Chlorophyceae

1. The members of Chlorophyceae are commonly known as grass green algae. The organization of the thallus varies widely. It ranges from unicellular, multicellular motile, non-motile, colonial, filamentous to complex thalloid forms.

2. Presence of pyrenoid and thus the reserve food is mainly in the form of starch but in siphonales it is in the form of oil drops.

3. Pyrenoids are present in chromatophores except in siphonales and charales.

4. Another important feature of this group is the occurrence of motile stages in the life cycle, motile bodies or swarmers contain two or more flagella of equal size attached on their front end.

5. Sexual reproduction ranges from isogamy to advanced oogamy through anisogamy.

ORDER	FAMILY	GENUS
1. Volvocales	1. Chlamydomonadaceae	Chlamydomonas
	2. Volvocaceae	Pandorina
		Eudorina
		Pleodorina
		Volvox
2. Chlorococcales	1. Chlorellaceae	Chlorella
	2. Hydrodictyaceae	Pediastrum
	3. Coelastraceae	Hydrodictyon
		Scenedesmus
3. Ulotrichales	1. Ulotrichaceae	Ulothrix
	2. Ulvaceae	Ulva
		Enteromorpha
4. Cladophorales	1. Cladophoraceae	Cladophora
5. Chaetophorales	1. Chaetophoraceae	Draparnaldiopsis
	2. Coleochaetaceae	Fritschiella
		Coleochaete
6. Oedogoniales	1. Oedogoniaceae	Oedogonium
7. Zygnematales	1. Zygnemataceae	Spirogyra
	2. Desmidiaceae	Zygnema
	3. Mesotaeniaceae	Cosmarium
		Netrium
8. Siphonales	1. Caulerpaceae	Caulerpa
		Codium
9. Charales	1. Characeae	Chara
		Nitella

CHLAMYDOMONAS: EHRENBERG, 1833

(Chlamydo - Clock, Moos - Single)

Systematic Position

Class   : Chlorophyceae

Order   : Volvocales

Family : Chlamydomonadaceae

Genus   : *Chlamydomonas*

Characteristic Features

1. The plant body is a thallus, which consists of a single biflagellated cell.

2. It is a microscopic, unicellular organism and it exhibits a very primitive type of structure.

3. They are ovoid, spherical, ellipsoidal or pyriform in shape.

4. *Chlamydomonas* cells are provided with two anterior flagella.

5. Each flagellum arises from a basal granule, the blepharoplast.

6. The two flagella or cilia propel the organism by their lashing movements in water.

7. The cavity of the cup-shaped chloroplast is completely filled with the cytoplasm in which lies a single nucleus.

8. Pyrenoids store reserve starch in the form of layers around it

9. The stigma is oval or circular and it is sensitive to light

10. A small projection of papilla, known as apical papilla, is present in between the two anteriorly inserted flagella.

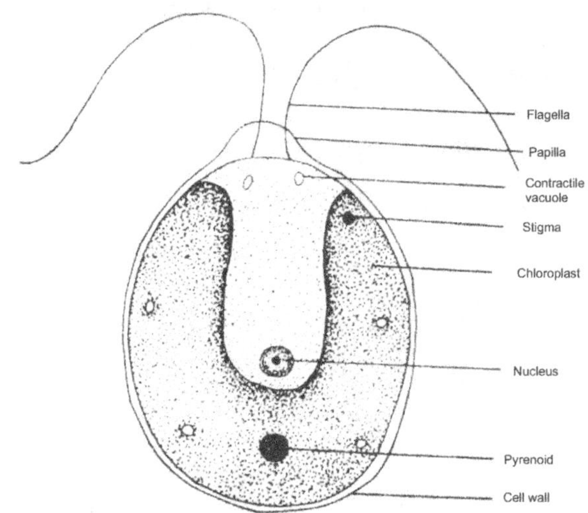

VOLVOX: LINNAEUS, 1758

(VOLVER - TO ROLL)

Systematic Position

Class   : Chlorophyceae

Order   : Volvocales

Family : Volvocaceae

Genus : *Volvox*

Characteristic Features

1. Colonies of Volvox are spherical to ellipsoidal in shape and they contain 500 to 50,000 cells.

2. Each cell is chlamydomonoid and is biciliate, motile and ovoid.

3. Each cell is provided with a gelatinous sheath of its own.

4. The cells are arranged in different manner and it is described as a coenobium.

5. Chloroplast is cup-shaped and a single nucleus is present.

6. Protoplast shows curved plate-like chloroplast with one or more scattered pyrenoids, and an eye spot is also present in cell.

7. *Volvox* exhibits asexual and sexual modes of reproduction.

8. The male reproductive cells are called androgonidial cells or antheridia. The female reproductive cells are called gynogonidia or oogonia.

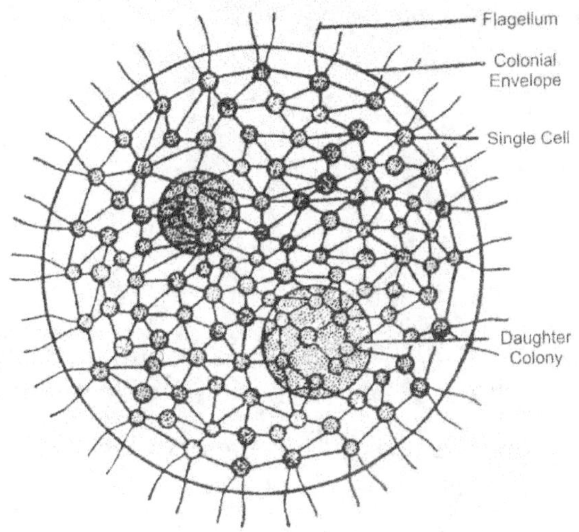

Volvox colony

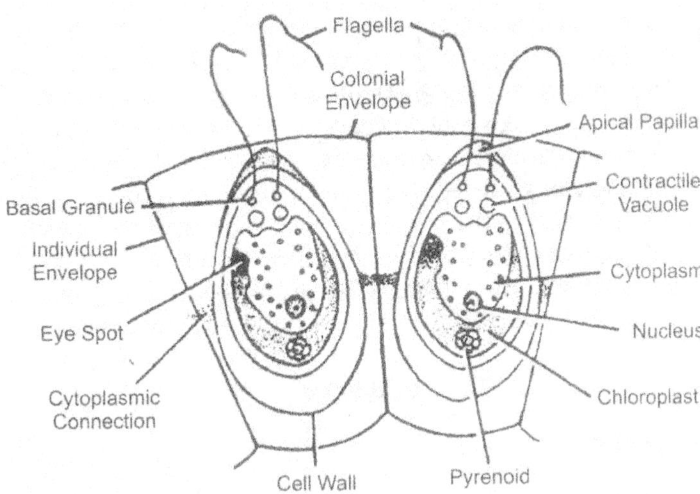

Volvox cell structure

CHLORELLA: Beijernick, 1890

(Chloro – green; ella – diminutive or affection)

Systematic Position

Class   : Chlorophyceae
Order   : Chlorococcales
Family : Chlorococcaceae
Genus   : *Chlorella*

Characteristic Features

1.   *Chlorella* is a unicellular and non-motile organism.

2.   It occurs on damp soils, barks of trees and fresh water pools and some species of chlorella are found in aquatic animals.

3.   Cells are spherical, ellipsoidal having cellulose cell wall and reserve food material is starch.

4.   Chloroplast is partially cup-shaped with or without pyrenoids. The single nucleus is in the colourless central cytoplasm.

5.   Zoochlorella is found in aquatic animal and freshwater sponges, these have the capacity to escape digestion.

6.   Only asexual reproduction is present.

7.   Multiplication takes place by autospores.

8.   *Chlorella* is used in many physiological experiments.

9.   *Chlorella* cells are rich in proteins, fats, carbohydrates and minerals; they may be used as food for human beings.

10. An antibiotic, chlorellin is extracted from chlorella.

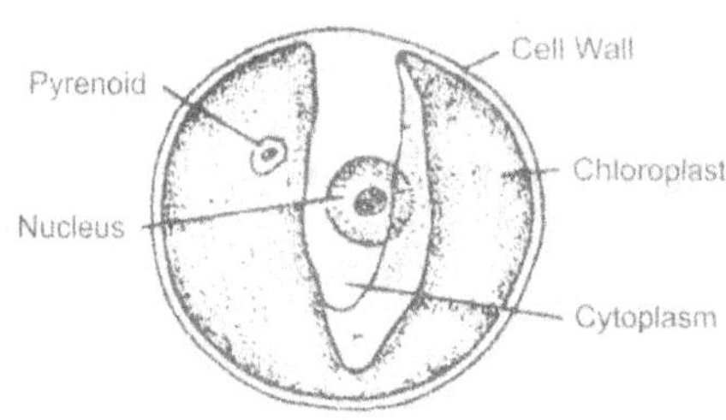

HYDRODICTYON: ROTH, 1880

Hydro - water: Diction - net

Systematic Position

Class   : Chlorophyceae

Order   : Chlorococcales

Family   : Chlorococcaceae

Genus   : *Hydrodictyon*

Characteristic Features

1. *Hydrodictyon* commonly known as "water-net" occurs in fresh water ponds and lakes.

2. The plant body is known as coenobium and a mature one may be as long as 20-30 cm.

3. The network of the coenobium is formed of a number of meshes.

4. The nets are generally delicate and they easily break up into individual cells.

5. In mature cells, the chloroplast is reticulate with many pyrenoids present in the cytoplasm.

6. Cell wall is formed of cellulose.

7. Large vacuole is present in the cytoplasm.

8. Both asexual and sexual reproduction is present. Asexual reproduction is by means of zoospores and sexual reproduction by means of repeated cleavage of protoplast produce biflagellate gametes.

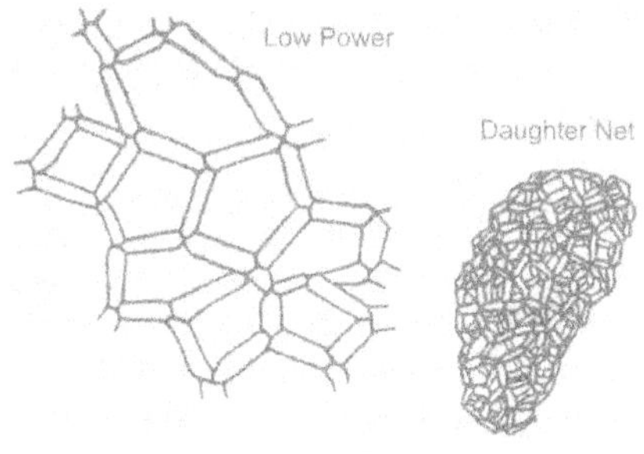

Hydrodictyon colony & daughter net

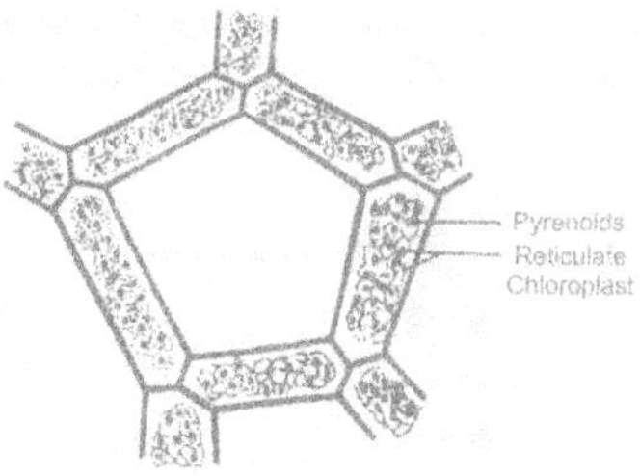

Single cell - Enlarged view

ULOTHRIX: Kuetzing, 1833

(Ulo – rough; thrix – hair)

Systematic Position

Class   : Chlorophyceae
Order   : Ulotrichales
Family : Ulotrichaceae
Genus   : *Ulothrix*

Characteristic Features

1. *Ulothrix* with about 30 species is mostly of freshwater form.

2. *Ulothrix* shows the simplest type of filamentous structure.

3. The cell wall is usually thin but in some species, it is thick and stratified.

4. All the cells are similar in structure and function except the basal rhizoidal cell.

5. The basal cell act as attaching organ and is usually deficient in chlorophyll.

6. Girdle-shaped chloroplast is present. Many pyrenoids are embedded in the chloroplast.

7. Reproduction takes place by three methods, vegetative, asexual and sexual methods.

8. Vegetative, asexual and sexual reproduction are by means of fragmentation, zoospores and gametes respectively.

9. The vegetative cell has a definite rigid cell wall, which is composed of inner cellulose layer and outer pectin layer.

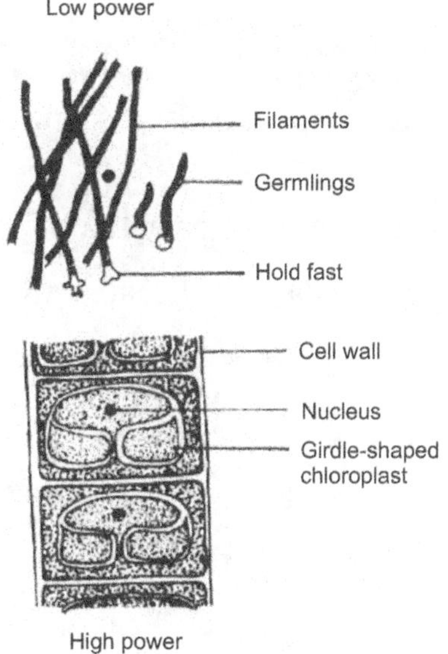

Ulothrix—Filaments in low power and cell magnified

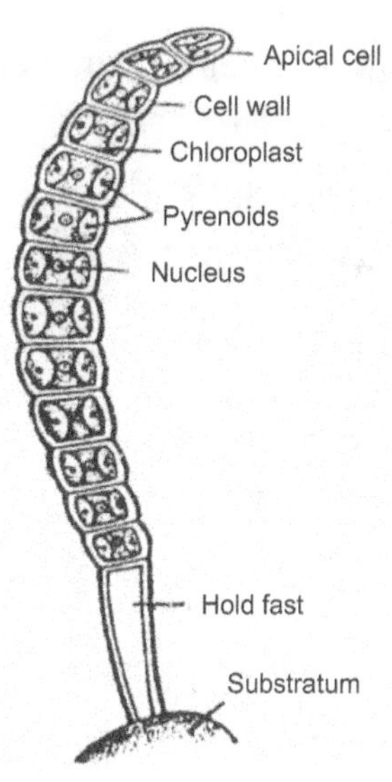

Ulothrix—Single filament enlarged

ULVA: Agarth, 1833

Systematic Position

Class : Chlorophyceae

Order : Ulotrichales

Family : Ulvaceae

Genus : *Ulva*

Characteristic Features

1. *Ulva* is an exclusively marine form, found attached to the rocks in the tidal zones of the oceans.

2. The thallus is an expanded leaf-like structure and two cells thick.

3. The basal part is like a narrow stalk expanded into a holdfast, it is pseudo-parenchymatous.

4. Each cell is uninucleate and isodiametric in shape.

5. Single parietal chloroplast is present with single pyrenoid embedded in the chloroplast.

6.   Cell division takes place always at right angles to the thallus surface.

7.   In the lower region, certain cells produce rhizoids.

8.   Certain cells of the basal part of the thallus produce some rhizoidal outgrowths.

9.   The rhizoids are multinucleate and produce secondary thalli.

10. Reproduction is by means of vegetative, asexual and sexual methods.

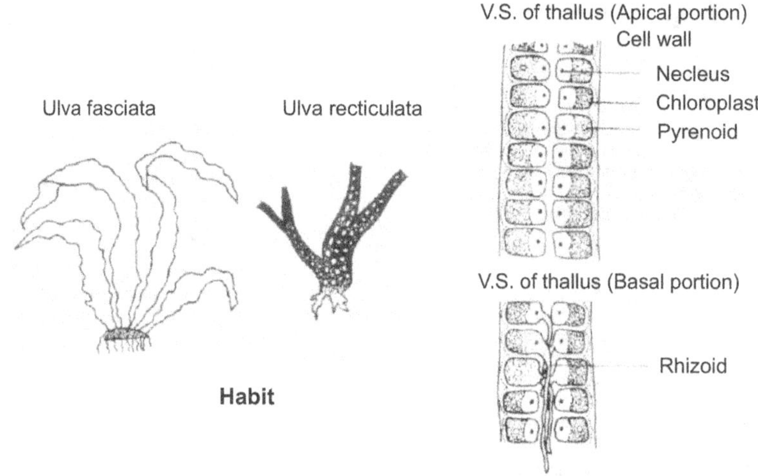

DRAPARNALDIA: Smith and Klyver

Systematic Position

Class   : Chlorophyceae

Order   : Chaetophorales

Family : Chaetophoraceae

Genus   : *Draparnaldia*

Characteristic Features

1.   It is a freshwater filamentous algae and is covered by a gelatinous sheath.

2.   The filament is attached to the substratum by means of branched multicellular rhizoidal branches, which are developed from the basal cell.

3. The thallus shows heterotrichy having a distinct erect system and poorly developed prostrate system.

4. The main axis is differentiated into small nodal cells and long internodal ells which alternate with each other in a regular pattern.

5. Nodal cell bear two types of branches, i.e., the long laterals and the short laterals.

6. Long lateral branches of unlimited growth and short lateral branches of limited growth are developed on main axis as well as in the long axis.

7. The apices of laterals usually end in pointed cells called setae.

8. The main axis is composed of large barrel shaped cells with a vacuolated cytoplasm and a single chloroplast with pyrenoid.

9. *Draparnaldia*, reproduce by asexual and sexual methods.

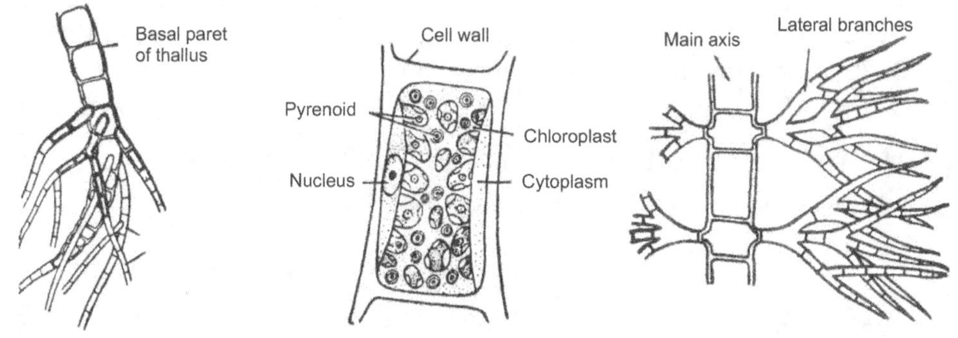

Draparnaldia - cell structure

Draparnaldia - habit

OEDOGONIUM: LINK, 1820

(Oedos - Swelling, Gonas - Fruit)

Systematic Position

Class   : Chlorophyceae

Order   : Oedogoniales

Family : Oedogoniaceae

Genus   : *Oedogonium*

Characteristic Features

1. *Oedogonium* is an unbranched multicellular filamentous freshwater form.

2. It is found epiphytic on submerged leaves and stems of hydrophytes.

3. The cell wall consists of three portions—the outer chitin, the middle pectin and the inner cellulose layer.

4. All the cells of the filaments are similar in shape except lowermost hold fast and uppermost apical cell.

5. Internal to cell wall is reticulate chloroplast with numerous pyrenoids present.

6. The pyrenoids accumulate starch in the form of plates.

7. Reproduction takes place by means of vegetative, asexual and sexual method.

8. Vegetative reproduction takes place by fragmentation, asexual method by zoospore and sexual reproduction by (male) antheridia and oogonia (female).

9. Mature and old cells show cap cells at their upper end, these are characteristic of the members of oedogoniales.

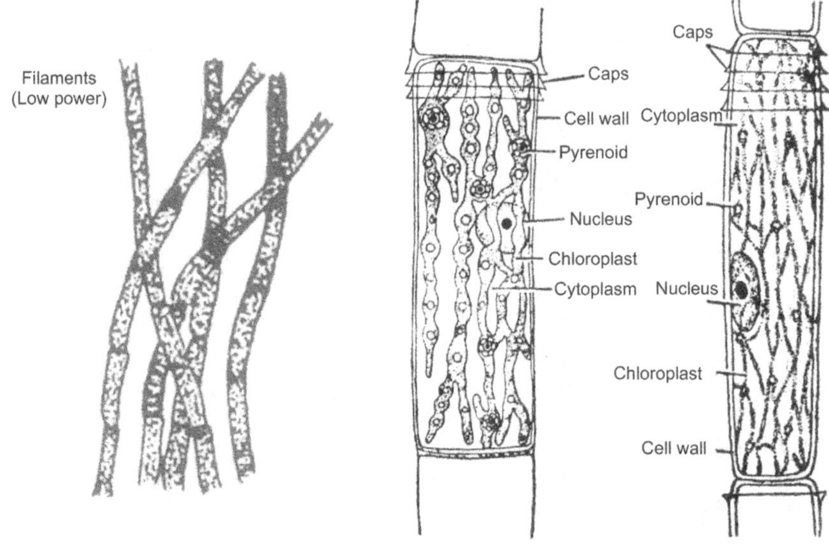

Oedogonium thallus   **Oedogonium - Cell structure**

Desmids: Bre'b

Systematic Position

Class   : Chlorophyceae

Order   : Conjugales

Family : Desmidiaceae

Genus   : *Desmids (Cosmarium)*

Characteristic Features

1. *Desmids* are essentially free floating forms, some may occur in terrestrial habitats.

2. Unicellular thallus consists of two hemispherical halves or semi cells.

3. A deep constriction present medially in each cell is known as sinus.

4. Two semi cells are joined by a bridge-like structure called the isthmus.

5. Cells are rod-like and do not show the two semi cells and isthmus.

6. Placoderm are true desmids, most of them are called semi cells. The gap between the semi cells is known as sinus.

7. The cell wall of placoderm desmids is formed of cellulose impregnated with iron and gelatinous sheath.

8. Chloroplast shows variation in structure and number.

9. Many pyrenoids are present in each cell.

10. Asexual reproduction is completely absent, sexual reproduction is by conjugation.

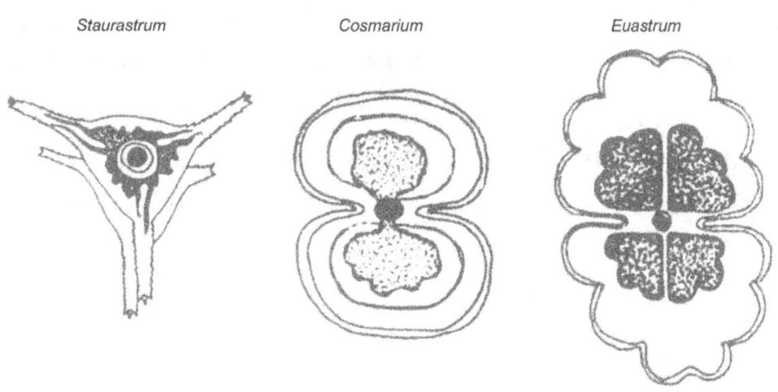

Caulerpa: Lamouroux, 1809

(Caul – stem; erpa - creep)

Systematic Position

Class   : Chlorophyceae

Order   : Siphonales

Family   : Caulerpaceae

Genus   : *Caulerpa*

Characteristic Features

1. *Caulerpa* is a marine algae with a multinucleate coenocytic thallus.

2. It is differentiated into three organs during its vegetative growth.

3. There is a cylindrical, rhizome-like, creeping portion.

4. From the upper surface of the horizontally growing rhizome arise a number of erect branches, which resemble foliage shoots and thus are often called the assimilatory shoots.

5. From the lower side of the rhizome arise the numerous, branched, thread-like colourless rhizoids.

6. Cross section of the thallus is coenocytic unseptate.

7. Form of the thallus is maintained by the turgor and thickness of the cell wall.

8. The cell wall is made up of cellulose, pectin, pectic acids and a polymer of pentose sugar.

9. There is no cellulose, the wall gradually increase in thickness behind the apex by successive deposition of these materials in successive strata.

10. Many transverse and longitudinal rods of callose and pectic material are present in the central region. These are known as trabeculae.

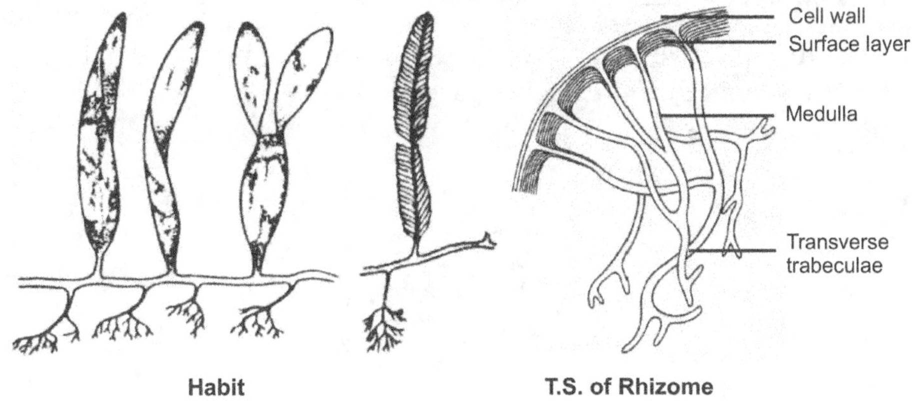

Acetabularia

Systematic Position

Class   : Chlorophyceae

Order   : Siphonales

Family   : Dasycladaceae

Genus   : *Acetabularia*

Characteristic Features

1. Siphonaceous or coenocytic thallus, the unicellular thallus is enlarged to form a nonseptate multinucleate sac-like or tubular structure which is not divided into cells in the somatic phase.

2. The former is tubular and the latter umbrella shaped.

3. *Acetabularia* consists of a stalk ending in an umbrella like cap.

4. The single nucleus lies at the base of the stalk.

5. The unicell reaches maturity and enters the reproductive phase. The nucleus undergoes division to form number of nuclei. This tendency to become multinucleate for a short time became, more permanent.

6. It is a tropical marine genus in which thallus extends to a foot or so in length.

7. It mimics the creeping shoots of aerial plants synergistically resembling a large moss or small angiosperm or a fern.

8.   It is differentiated into a creeping structure resembling rhizome.

9.   The rhizome give rise to root like holdfast from its under face and erect leaf shoot from its upper face.

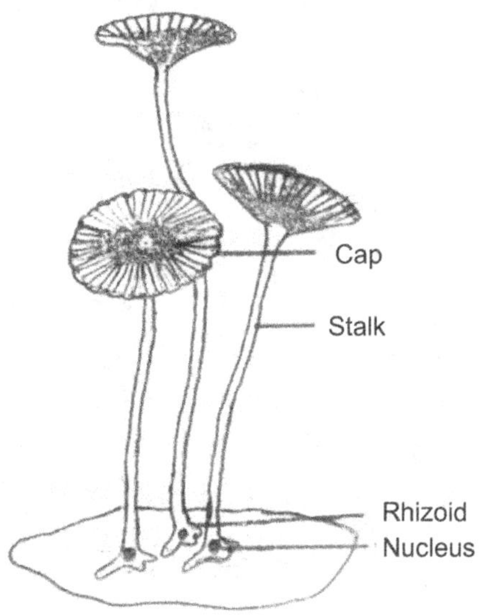

HALIMEDA

Systematic Position

Division : Chlorophyta

Class   : Chlorophyceae

Order   : Siphonales

Family   : Codiaceae

Genus   : Halimeda

Characteristic Features

1.   The disc structure is most highly perfected in *Halimeda* in which the branched thallus is composed of a number of usually flat (often cordate or reniform) strongly calcified segments.

2.   These are separated by uncalcified joints and arise from a short basal stalk attached to the creeping system.

3.   *H. tuna* is slightly calcified occurring commonly in the Mediterranean. Species of Halimeda may play a considerable role in formation of some coral-reefs.

4.   Internally, plant body is characterized by the complete absence of septation and is composed of closely apposed and intertwined coenocytic thread traversed by longitudinal and transverse skeletal strands.

5.   The plant body is a single multinucleate cell with a central vacuole and lining cytoplasm with numerous discoid chloroplasts.

6.   Carbonate of lime is deposited as a continuous layer on the inner surfaces of the facets, the resulting cylinder being perforated only by narrow pores traversed by the slender stalks of the secondary branches.

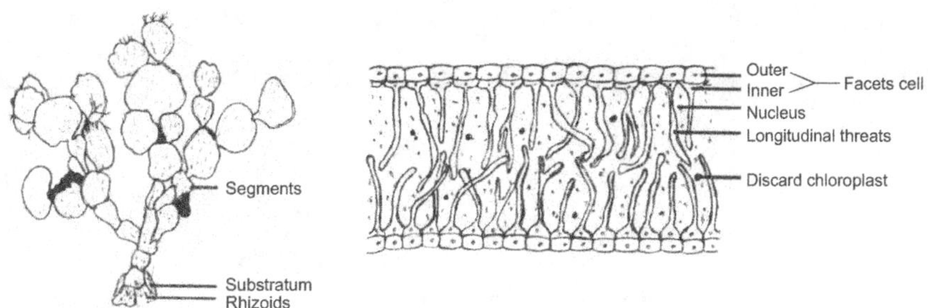

Codium

Systematic Position

Division : Chlorophyta

Class   : Chlorophyceae

Order   : Siphonales

Family   : Codiaceae

Genus   : *Codium*

Characteristic Features

1.   *Codium* occurs commonly in deeper rock pools with the plants being firmly attached by their basal discs and floating out into the water or hanging down from the rocks at the low tide mark on the rocky ledge.

2.   Plants are dark green in colour, often much elongated, 10-30 cm high, 3-5 mm diameter, and regularly dichotomous.

3. Younger divisions of the thallus are terete while older ones are frequently compressed and slightly expanded beneath furcations.

4. Utricles are obovate and clavate giving 1-4 coenocytic filaments at the base.

5. Sporangia are one or more on each utricle, sometimes the utricles bear scars of previously borne sporangia.

6. Sporangia are stalked, lanceolate, pointed, and broad with a long opening through an apical pore.

7. The transverse section of thallus is composed of a central medulla of narrow, interwoven, forked threads and a peripheral cortex of large club-shaped vesicles densely grouped at the same level to form a palisade layer.

8. Chloroplast is aggregated beneath the outer surface of each cortical vesicle.

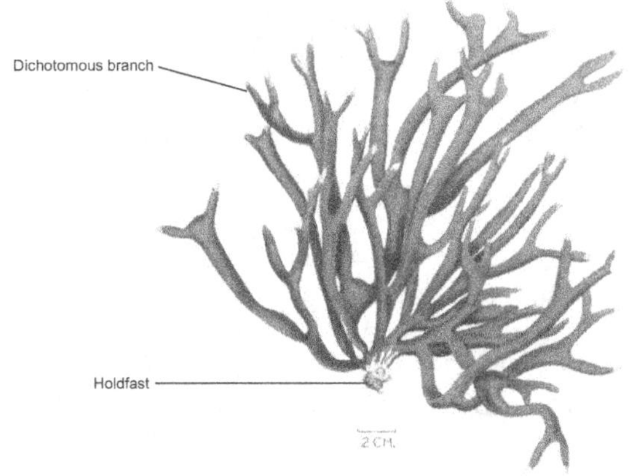

VALONIA

Systematic Position

Division : Chlorophyta
Class   : Chlorophyceae
Order   : Siphoncladales
Family   : Valoniaceae
Genus   : *Valonia*

Characteristic Features

1. It is multicellular in character, but basically a multinucleate structure.

2. The distribution is more in tropical and subtropical seas, although a few are found in the Mediterranean.

3. It is a macroscopic young plant, often with a club shaped vesicle anchored basally by rhizoids of various types and possessing the usual siphoneous structure.

4. Each and every vesicle is aggregated to one another and possesses a nucleus. Each vesicle has a balloon like structure.

5. The lobed chloroplasts are characteristically arranged in a network.

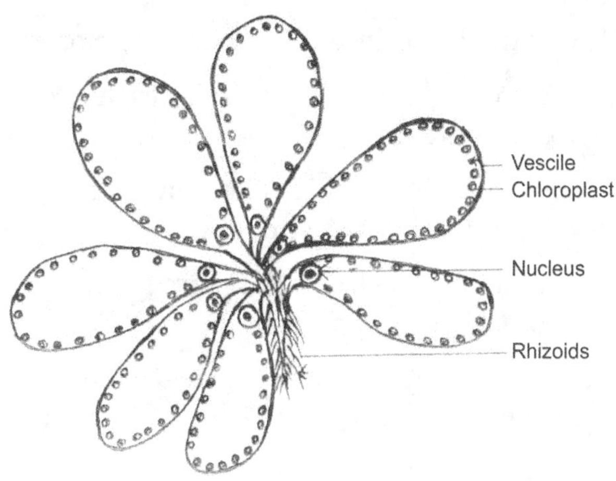

CHAROPHYCEAE

CHARA: VALLIANT, 1719

(Stone Wort)

Systematic Position

Class : Charophyceae

Order : Charales

Family : Characeae

Genus : *Chara*

Characteristic Features

1. The genus *Chara* has about 90 species that are widely distributed in standing waters.

2. Mostly, they are fresh water forms, a few are marine forms.

3. The plant is attached to the muddy bottom of the pond or pool by means of rhizoids.

4. Rhizoids are thread like, uniseriate, multicellular, well-branched and white coloured structures.

5. These arise from the lower nodes of the main axis.

6. Oblique septa are present in rhizoids.

7. The plant is distinguished into internodes and nodes.

8. The internode is constituted by long cylindrical, cortical cells which are much smaller in diameter.

9. At the nodes of the leaves, single-celled branches occur and they are known as secondary laterals or stipules.

10. Single nucleus and several chloroplasts are present in the dense cytoplasm, chloroplast lacks pyrenoids.

11. *Chara* reproduces by vegetative and sexual methods. Asexual reproduction is entirely absent.

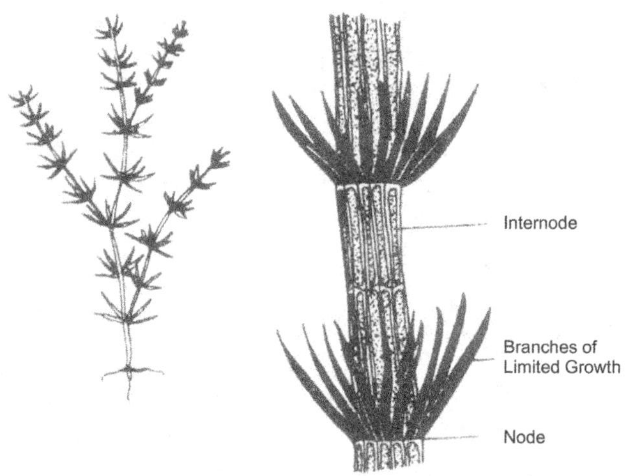

NITELLA: AGARDH, 1824

(A little star)

Systematic Position

Class	: Charophyceae
Order	: Charales
Family	: Characeae
Sub Family	: Nitelloideae
Genus	: *Nitella*

Characteristic Features

1. *Nitella* occurs in shallow water along the edges in pools, lakes and slow flowing streams attached to the sandy or muddy substratum.

2. Macroscopic branched thallus give the appearance of a small equisetum plant. Nitella has a jointed, central main axis with a whorl of branches arising from each joint. It is anchored to the substratum by multicellular, branched, and colourless rhizoids.

3. The main axis is differentiated into a series of well-marked nodes and internodes. The internode is a single, elongated, individual cylindrical cell several times longer than broad without an ensheathing layer of cortex (ecorticate).

4. The more fluid, less denser endoplasm is in a state of constant rotation. This streaming movement of endoplasm is termed cyclosis.

5. The main axis and the auxillary branches show indefinite growth by the activity of a dome shaped apical cell located at the tip.

6. Sexual reproduction is oogamous. The sex organs are large, highly specialized and complicated in structure.

7. Antheridium consists of a jacket layer of eight closely fitting shield cells. From the middle of each arises elongated cell called the manubrium.

8. The manubrium at its distal end bears a terminal cell known as primary capitulum which buds about six secondary capitula.

9.   The diploid zygote or the oospore germinates after a resting period to produce the haploid protonema.

10.   The *Nitella* plant is thus haploid; the only diploid structure in the life cycle is the dormant zygote.

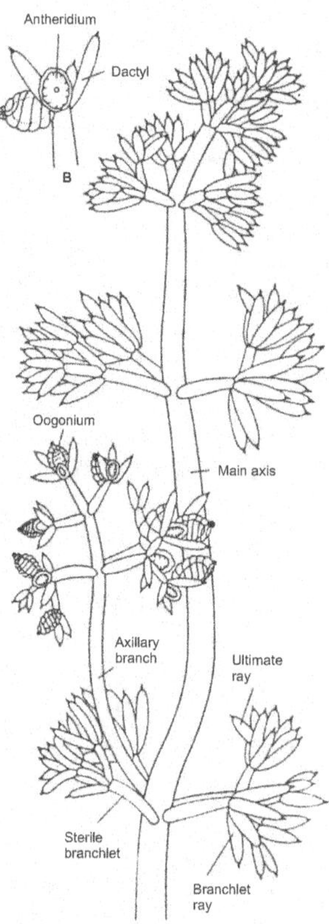

CHROMOPHYTA

XANTHOPHYCEAE

General Characteristic Features of Xanthophyceae

1.   Cell wall is made up of two layers, the main components of which are pectic compounds.

2.   Important pigments are chlorophyll a, chlorophyll e, β-carotene and xanthophylls like lutein, violaxanthin and neoxanthin.

3. Reserve food materials are fat or oils, starch is absent and pyrenoids are very rarely present.

4. Motile bodies contain two unequal flagella.

5. Chromatophores are discoid in shape.

Classification

$$Xanthophyceae \rightarrow$$

→ Heterococcales - coccoid forms
→ Heterochloridales - motile forms
→ Heterotrichales - filamentous forms
→ Heterosiphonales - siphonaceous forms

VAUCHERIA: De comdole

Systematic Position

Class : Xanthophyceae

Order : Heterosiphonales

Family : Vaucheriaceae

Genus : *Vaucheria*

Characteristic Features

1. It is commonly found in fresh water ponds or on moist soil in the form of green mat.

2. The plant body consists of a well branched siphon like, coenocytic filaments, i.e., the filaments are aseptate, multinucleate continuous structures.

3. Terrestrial species are attached to substratum by means of a branched hyaline cell known as holdfast or hapteron.

4. The cell wall is very thin, consisting of outer pectose and inner cellulose layers.

5. Reserve food material is present in the form of oil drops, lipid and leucosin or chrysolaminarin pyrenoids are absent.

6. The nuclei are present in the cytoplasm towards the central vacuole.

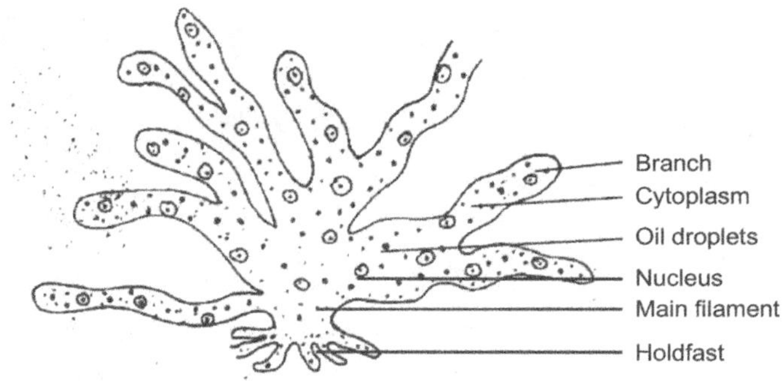

Botrydium: Wallroth, 1815

(a small cluster)

Class	: Xanthophyceae
Order	: Heterosiphonales
Family	: Botrydiaceae
Genus	: *Botrydium*

External Features

1. It is a unicellular, coenocytic, terrestrial alga consisting of an aerial and an underground portion.

2. The aerial portion is a spherical or pear shaped vesicle, measuring 1-2 mm, in diameter.

3. Plant body is differentiated into an underground rhizoidal portion and an aerial or subaerial vesicular portion.

4. Vesicle is spherical or balloon shaped and green in color but in the shady conditions the vesicle becomes elongated.

5. Rhizoidal portion is well branched, underground, colourless and aseptate.

6. Pyrenoids are present when the vesicles are young but whether these bodies are present in the mature condition also is not clear.

7. Reserve food material in the form of oil or leucosin starch is totally absent.

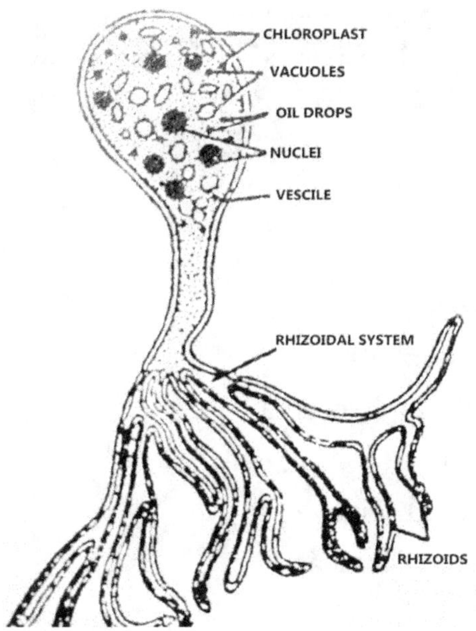

BACILLARIOPHYCEAE

Important Characteristic Features are:

1. Widely distributed in sea, freshwater and soil.

2. Chromatophores are yellow or golden yellow in colour.

3. Reserve food material is in the form of oil or volutin granules.

4. Pyrenoids are present but lack starch sheath.

5. All the members are unicellular but form colonies when many members come together.

6. Cell wall is made up of pectic material and silica.

7. The cell wall consists of two halves arranged in the manner of a box and lid. The outer half is known as epitheca and the inner half as hypotheca.

8. Members are either radially symmetrical (order Centrales) or bilaterally symmetrical (order Pennales).

9. Centrales: Radially symmetrical, raphe absent. Many chromophores per cell, reproduction by auxospores.

10. Pennales: Symmetrical or even asymmetrical, presence of raphe, one or two chromatophores per cell, reproduction by auxospores and statospores.

Cyclotella

Systematic Position

Phylum   : Chrysophyta

Division   : Chromophyta

Class   : Bacillariophyceae

Order   : Centrales

Suborder   : Coscinodiscineae

Family   : Thalassiosiraceae

Genus   : Cyclotella

Characteristic Features

1. Cells solitary, valves tangentially undulate and valve wall is alveolated.

2. A central field distinctly different from the rest of the valve.

3. Central portion with pflexes and coarsely punctate.

4. Cells in chains fairly close, separated by occluded process on marginal part valve.

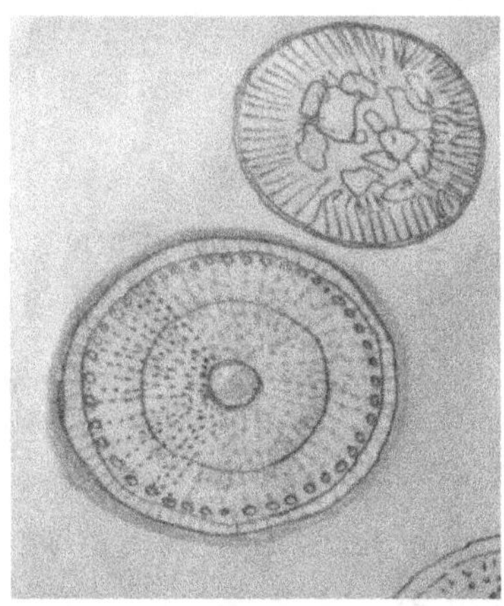

Pinnularia

Systematic Position

Phylum : Chrysophyta

Division : Chromophyta

Class   : Bacillariophyceae

Order   : Pennales

Family   : Naviculoideae

Genus   : Pinnularia

Characteristic Features

1. It is a unicellular, microscopic alga, which is commonly known as diatom.

2. Each cell is oval or elliptical in shape with bilateral symmetry.

3. Each cell consists of a frustule, which is composed of two overlapping halves (theca), the epitheca and hypotheca.

4. Epitheca overlaps the hypotheca and the overlapping portion is known as girdle or connecting bond.

5. Striations are arranged isobilaterally, i.e., on either side of the central line.

6. The central line has a longitudinal slit which runs from one polar nodule to the other and is known as raphe.

7. The raphae has two polar nodules and a central nodule. These nodules are the internal thickenings of the wall.

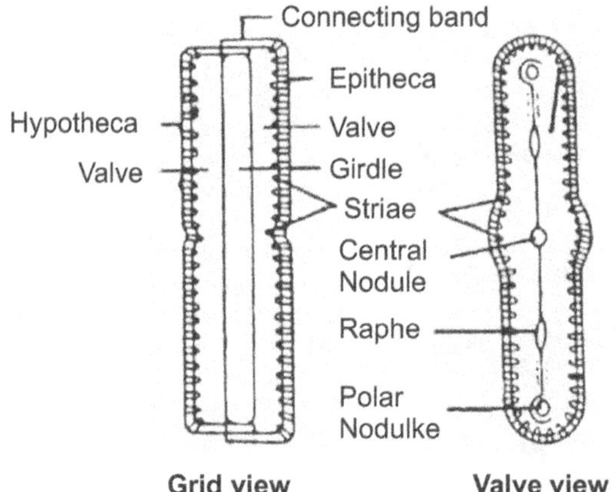

PHAEOPHYCEAE

Salient Features of Phaeophyceae

1. Pigments are chlorophyll a and c, β-carotene, flavoxanthin, fucoxan-thin, violaxanthin and luetin.

2. Due to the presence of fucoxanthin, the members are brown in colour and are commonly called brown algae.

3. Reserve food materials are mannitol and laminarian. Pyrenoids are present only in some lower forms.

4. Reproduction units are motile having two flagella attached laterally.

5. Motile reproductive bodies are developed either in unilocular or pluriloc-ular sporangia.

6. Sexual reproduction ranges from simple isogamy to oogamy.

The class includes nine orders:

1. Ectocarpales

2. Tilopteridales

3. Cutleriales

4. Sporochanales

5. Desmarestiales

6. Laminariales

7.   Sphacelariales

8.   Dictyotales

9.   Fucales

Anatomical Features

1.   The outermost covering layer on both sides in a single celled thick meristoderm or epidermis consists of small rectangular cells. They are uninucleate, full of chromatophores and help in photosynthesis.

2.   Meristoderm is followed by central medulla which is composed of large rectangular cells with few or no chromatophores. These cells are meant for storage of reserve food and contain fucosan vesicles and refractive granules, etc.

3.   Multicellular hair in the form of tufts may be observed on both the surface of thallus.

4.   In between meristoderm and medulla few layers of cortex may be perfect sometimes medulla is many celled thick.

5.   Asexual reproduction takes place by means of tetraspores, sexual reproduction is oogamous and the plants are strictly dioecious.

Ectocarpus

Systematic Position

Division   : Phaeophyta

Class   : Phaeophyceae

Sub Class   : Isogeneratae

Order   : Ectocarpales

Family   : Ectocarpaceae

Genus   : *Ectocarpus*

Characteristic Features

1.   This marine alga is the most primitive of all the brown algae and comprises many species and a few have been reported to occur in freshwater.

2.   The thallus is fine, tufted, brownish heterotrichous filament consisting of well defined erect and prostrate system.

3.   The filaments are generally uniseriate and usually consist of a single stand of cells.

4.   The branching is always lateral, end in colourless mucilage hairs.

5.   The cells are short and cylindrical; cell wall is made up of pectic-cellulose. The protoplasm consists of golden brown disc or band shaped chromatophores with characteristic pigment fucoxanthin.

6.   Genetically the thalli of Ectocarpus are of two kinds, haploid and diploid. Morphologically they are alike.

7.   The haploid thalli are concerned with sexual reproduction and diploid with asexual reproduction.

8.   The diploid plants bear two kinds of sporangia.

    i.   Unilocular produce haploid zooids or swarmers called meiozoospores.

    ii.   Plurilocular gives rise to single diploid zoospore.

9.   Sexual reproduction is isogamous, takes place by the formation of biflagellate gametes structurally similar to the zoospores produced in gametangia.

10.   The haploid and diploid plants are morphologically alike. Ectocarpus thus exhibits an isomorphic type of alternation of generations.

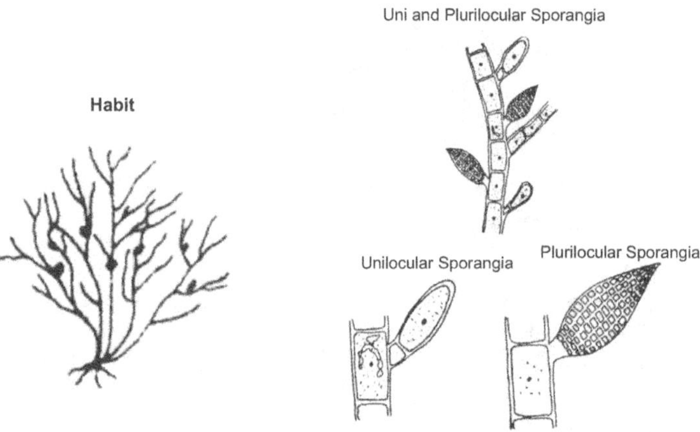

Unit and Plurilocular Sporangia

Padina

Systematic Position

Division : Phaeophyta

Class : Phaeophyceae

Order : Dictyotales

Family : Dictyotaceae

Genus : *Padina*

Characteristic Features

1. *Padina* is a marine alga. The thallus consists of stalk and fan shaped fronds, 5-12 cm in height; the larger ones often loosely rolled on their longitudinal axis like a cornet are distinctive of all the species of the genus.

2. The stalk is the upward continuation of the prostrate rhizome.

3. The rhizome is very much branched and attached to the substratum by tufts of rhizoids.

4. The thallus shows concentric zones which are arranged in 4-8 rows. They are formed by the hairs.

5. The zones of hairs occur alternately on the two surfaces but are more strongly developed on the upper and are stated to be specially prominent in plant exposed to strong light.

6. The thallus shows an apical cell and row of marginal initials. The older portion of the thallus has the maximum of 10 or more cells in thickness.

7. The cells are more are less isodiametric and with iodine.

8. The longitudinal section of the thallus from the apex to the interior shows circinate coiled vernation.

9. These cells are in the protection of the apical cells.

10. On the thallus surface, trichomes and sporangia or gametangia of two kinds are found on the older parts. The gamete producing organs are known as gametangial sours.

Padina habit

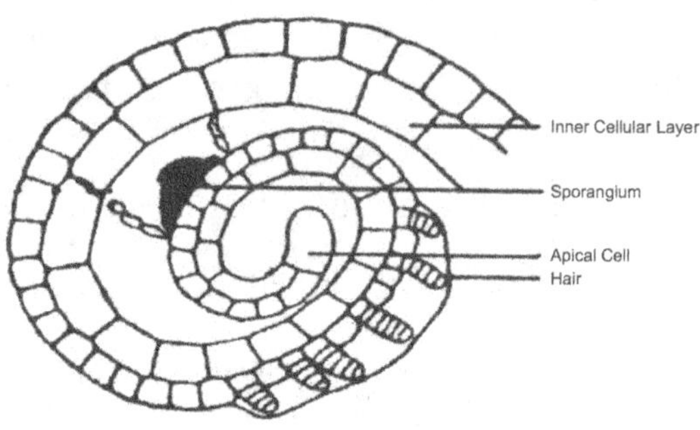

Padina thallus L.S.

Dictyota: Lamouroux, 1809

Systematic Position

Class   : Pheophyceae

Order   : Dictyotales

Family : Dictyotaceae

Genus   : *Dictyota*

General Characteristics

1. Well developed and branched thallus is attached to the substratum by a disc like holdfast.

2. Holdfast is prostrate and irregular in shape.

3. From the holdfast arise many erect, branched leafy structures or fronds.

4. A frond consists of a lower cylindrical stalk and an upper well branched portion.

5. Fronds are flattened structures and lack a midrib.

6. At the apex of each branch is present an apical cell. Apical cell divides into two, thus developing into two branches of dichotomy.

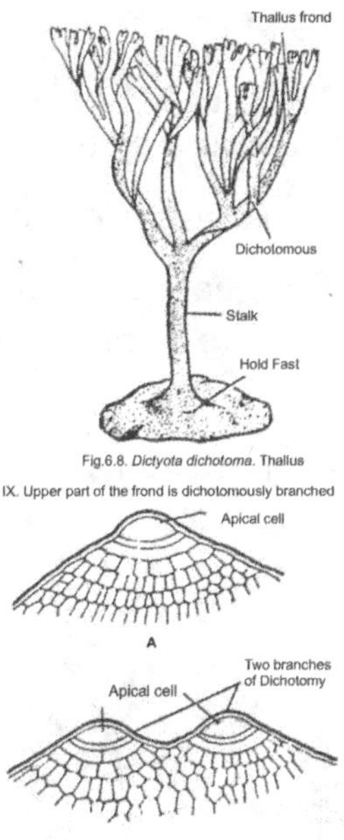

Fig.6.8. *Dictyota dichotoma*. Thallus

IX. Upper part of the frond is dichotomously branched

Stoechospermum

Systematic Position

Division : Phaeophyta

Class : Phaeophyceae

Order : Dictyotales

Family : Dictyotaceae

Genus : *Stoechospermum*

Characteristic Features

1. Thallus is in the form of large tufts, 15-25 cm or more high.

2. Base with irregularly ramified, bent, decumbent branches interwoven.

3. Rhizoids numerous from the base, attaching the plant to the substratum.

4. Thallus erect, spatulate, dichotomously branched repeatedly a few times, flat, without midrib; segments of thallus 1-2 cm broad; margin entire; apex bifid or flatly truncate.

5. Section of thallus shows greater part with large parenchymatous cells in the middle and on either side covered by two layers of small cells.

6. Hairs in groups scattered on both sides of thallus.

7. Tetrasporangia in longitudinal rows in irregular outline near the margin and extending to some distance along the margin of the thallus, 0.2-0.3 cm thick.

8. Thallus brown in younger parts, tending to be dark brown in the older and basal part of the thallus.

Sargassum

Systematic Position

Division : Phaeophyta (Phaeophycophyta)

Class   : Phaeophyceae

Order   : Fucales

Family   : Sargassaceae

Genus   : *Sargassum*

Characteristic Features

1. *Sargassum* is a marine alga popularly called the gulf weed, occurs in tropical or temperate zones attached along the rocky shores.

2. It is macroscopic and more or less bushy in habit, looks like a small angiospermic plant found attached to the substratum by means of well developed holdfast.

3. The thallus is mainly composed of the much branched primary, long laterals.

4. The lower members of the auxillary branch system may be reduced to air-bladders. The subsequent ones are fertile and may be cylindrical to flattened called receptacles.

5. It reproduces entirely by fragmentation and sexual reproduction.

6. Sexual reproduction is oogamous. The sex organs are lodged in conceptacles which are confined to the receptacles.

7. Some species are monoecious and others dioecious.

8. The zygote germinates without undergoing any resting period and forms the diploid thallus.

9. *Sargassum* is a source of alginates and alginic acid.

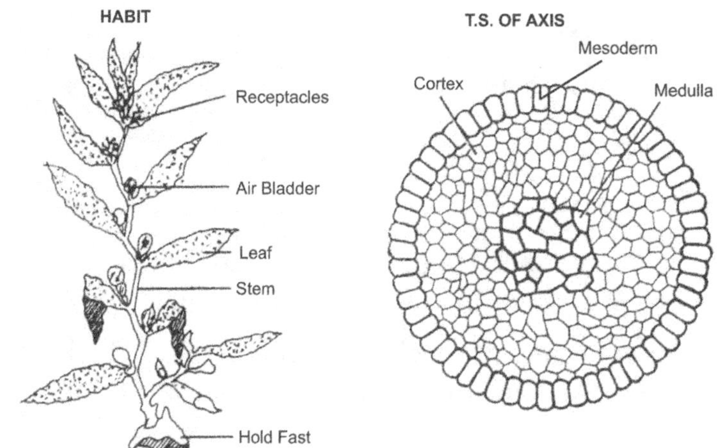

TURBINARIA

Systematic Position

Division : Phaeophyta

Class : Phaeophyceae

Order : Fucales

Family : Sargassaceae

Genus : *Turbinaria*

Characteristic Features

1. *Turbinaria* is a multicellular marine alga.

2. It is found round the year on the rocks, stones and old corals in the sub tidal region. It occurs only in less quantity.

3. The thallus is yellowish brown or brown in colour, 20-25 cm in height, possess elongated cylindrical branches, leaves possessing angular margins.

4. It is used as raw material for the manufacture of sodium alginate and also yields 8-10% of mannitol.

Turbinaria conoides

Turbinaria sps.

RHODOPHYTA

RHODOPHYCEAE

General Characteristics Features of Rhodophyceae

1. Red algae are eukaryotic cells containing pigments chlorophyll a, β-carotene and γ-phycoerythrin, chlorophyll d, alpha-carotene, lutein and γ-phycocyanin.

2. Food reserve may occur as alcohols but are chiefly stored in the form of polysaccharides as small granules free in the cytoplasm rather than in plastids and are known as florridean starch.

3. The occurrence of non-aggregated photosynthetic lamella or thylakoids with the plastids.

4. Growth of the thallus may be apical as well as intercalary; one of the striking features of red algae is the conspicuous cytoplasmic connection from cell to cell.

5. Outstanding characteristic of this taxon is a total absence of any flagellated reproductive cell.

6.   Sexual reproduction is of most advanced oogamous type.

7.   Male sex organs are known as spermatangia and female as carpogonia. Carpogonium is a flask shaped structure having a long neck like trichogyne.

Classification

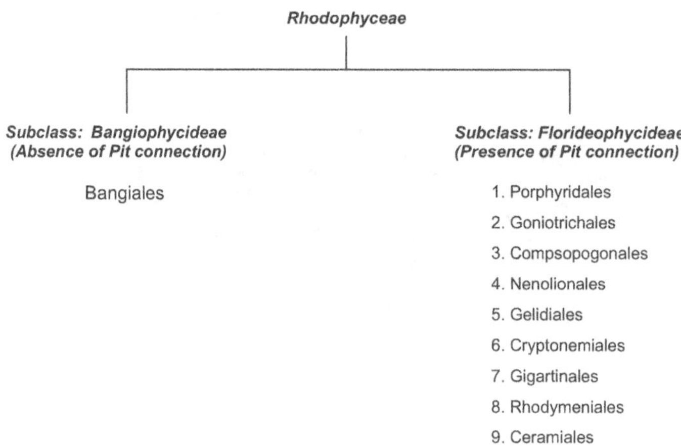

BATRACHOSPERMUM: ROTH, 1797

Batracho - Frog, Spermum - Seed

Systematic Position

Division   : Rhodophyta (Rhodophycophyta)

Class   : Rhodophyceae

Sub Class   : Florideae

Order   : Nemalionales

Family   : Batrachospermaceae

Genus   : *Batrachospermum*

Characteristic Features

1.   It is one of the few, inland fresh water red algae commonly found in tropical, sub tropical and temperate regions.

2.   It occurs in slow flowing streams, water falls, margins of lakes, in springs usually under the shade attached to rocks.

3.   The thallus is bluish or olive green or violet green in colour when the plants are growing in shallow water and dark violet or reddish colour in deep water specimens.

4.   The adult thalli are uniaxial in its simplest form, heterotrichous, may attain a length of 20 cm or even more.

5.   The branches (or) laterals it bears are of two kinds:

     i.   Primary laterals or branches of limited growth

     ii. Long branches or branches of unlimited growth

6.   The growth of axial filament and branches takes place by means of apical cell.

7.   The sexual reproduction is of an advanced oogamous type by formation of male sex organs antheridia or spermatangia and female carpogonia.

8.   The zygote developed as a result of fertilization undergoes meiosis to form haploid gonimoblast filament, which in turn gives rise to cystocarp or carposporophyte on parent.

9.   The cystocarp produces a haploid carpospore, germinate to form protonema like heterotrichous structure called chantransia stage and later develops into adult *Batrachospermum*.

10. Asexual reproduction is brought about by means of non-motile, uninucleate spores called monospores.

11. Cytologically the life cycle of *Batrachospermum* is called haplobiontic and morphologically as triphasic or trigenic or haplohaplohaplontic

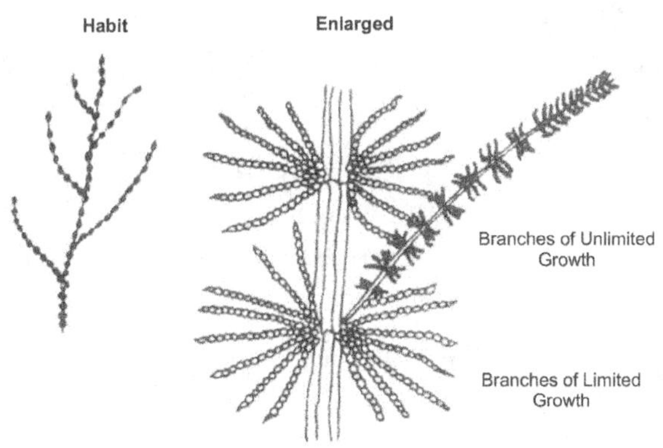

POLYSIPHONIA: GREVILLE, 1824

(Poly – Many, Siphonia - Siphons)

Systematic Position

Division	:	Rhodophyta (Rhodophycophyta)
Class	:	Rhodophyceae
Sub Class	:	Florideae
Order	:	Ceramiales
Family	:	Rhodomelaceae
Sub family	:	Polysiphonieae
Genus	:	*Polysiphonia*

Characteristic Features

1. The plant body is polysiphonous in nature consisting of the basement and the upright portion.

2. The feathery, upright portion consisting of a much branched system of relatively large branches bearing numerous small branches is called the trichoblast. The later bears the sex organs.

3. Occurrence of pit connections between the adjoining central cells and between the central cells and the pericentral cells is the characteristic feature of red algae.

4. Occurrence of three kinds of plants—gametophytes, carposporophyte and tetrasporophyte.

5. Sexual reproduction is oogamous. The male sex organ is called the spermatangia and the female sex organ is called carpogonia (flask shaped).

6. The diploid nucleus formed by fertilization lies at the basal swollen portion of the carpogonium. The latter communicates by means of a tubular connection with the auxillary cell cut off from the supporting cell.

7. The development of gonimoblast filaments forms the auxillary cell.

8. The formation of a diploid carpospore in the modified terminal cell of the gonimoblast filament is now called the carposporangium bearing the carpospores.

9. Development of the tetrasporophyte by the germination of the diploid carpospore, is concerned with production of meiospores called tetraspores (haploid).

10. Germination of haploid tetraspores gives rise to the gametophyte or sexual plants concerned with sexual reproduction.

11. The life cycle with three phases in which one is haploid and two diploid is called diplodiplohaplontic or diplobiontic.

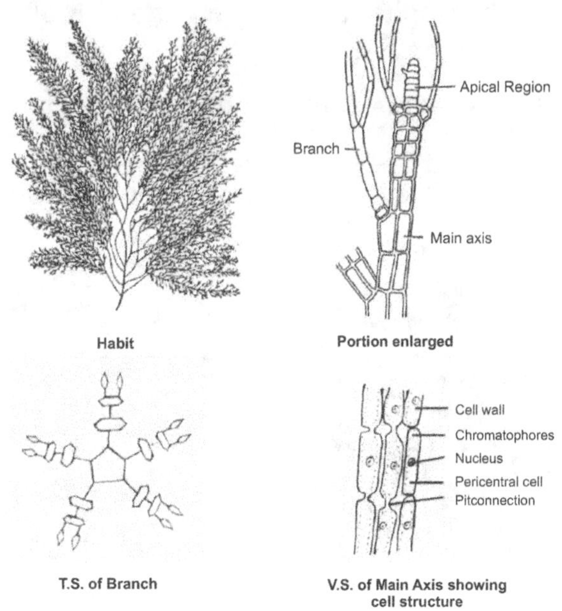

Gracilaria

Systematic Position

Division : Rhodophyta

Class : Rhodophyceae

Order : Gracilariales

Family : Gracilariaceae

Genus : *Gracilaria*

Characteristic Features

1. Plants are attached to the substratum by the thin spreading disc.

2. Fronds are 5-10 cm high, 2-4 mm wide, compressed, except near the base where it is subterete.

3. The branches given off from the edges of the flat thallus are either alternate or seriate (with 2-3 series) from the same side.

4. The upper parts thus get an antler like appearance up to 4 mm broad below the ramification, apex acute or bifid.

5. Medullary tissue of large cells are up to 250 μm diameter

6. Cystocarps are hemispherical, prominent and profusely scattered on the flat side of the thallus.

7. Spores are rounded and of cartilaginous consistency

8. The transverse section of the thallus shows the outer 1-2 layers of epidermis, middle 2-3 layers cortex and inner 1-2 layered medulla.

Habit

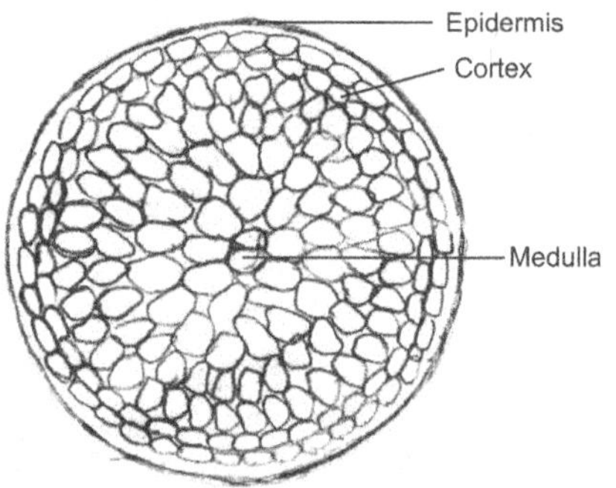

T.S. of Thallus

Hypnea

Systematic Position

Division : Rhodophyta

Class   : Rhodophyceae

Order   : Gigartinales

Family   : Hypneaceae

Genus   : *Hypnea*

H. musiformis - Vegetative Plants

H. musiformis – Vegetative Plants

Characteristic Features

1. Thallus shows a uniaxial construction.

2. The outer cortical region consists of a single layer of small densely pigmented cortical cells. Each cell is about 10-25 mm diam.

3. The subcortical region consists of 1-2 layers of cells. Cells are 30-50 mm in diam.

4. The medullary region consists of 2-3 layers of larger cells. Each cell is measuring about 100-300 mm diam.

5. The cortical, subcortical and medullary cells are progressively large in size towards the centre. At the centre a distinct axial cell is present which is about 25-40 mm diam.

6. The anatomical features of the vegetative portions of cystocarpic and tetrasporic plants are similar to that of the vegetative plants.

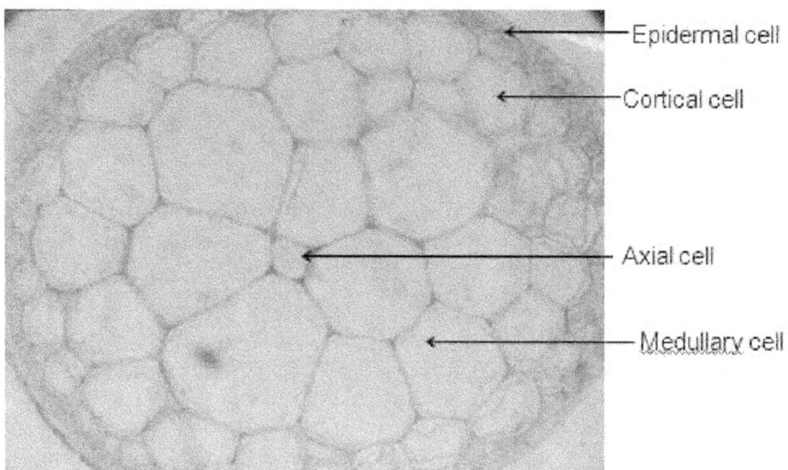

Cross section of **Hypnea**

CYANOPHYCEAE

General Characteristic Features of Myxophyceae (Cyanophyceae)

The members of this class are commonly known as blue green algae. Recently this group is also called as cyanobacterium because of characteristics similarities with bacteria.

1. The chief pigments present in this group are C-phycocyanin and C-phycoerythrin besides other usual pigments (e.g., chl-a, b-carotene, flavacin, etc.).

2. The cell is prokaryotic in nature; means no definite nucleus, chromosome and subcellular organelles but nuclear material is present in the form of DNA.

3. Photosynthetic reserve food substance is cyanophycin starch.

4. Cell wall is made up of mucopeptide or muramic acid.

5. Vegetative reproduction is by means of fission, fragmentation, hormogonia, akinetes, endospores, exospores and nanocyst. Sexual reproduction is completely absent.

Classification

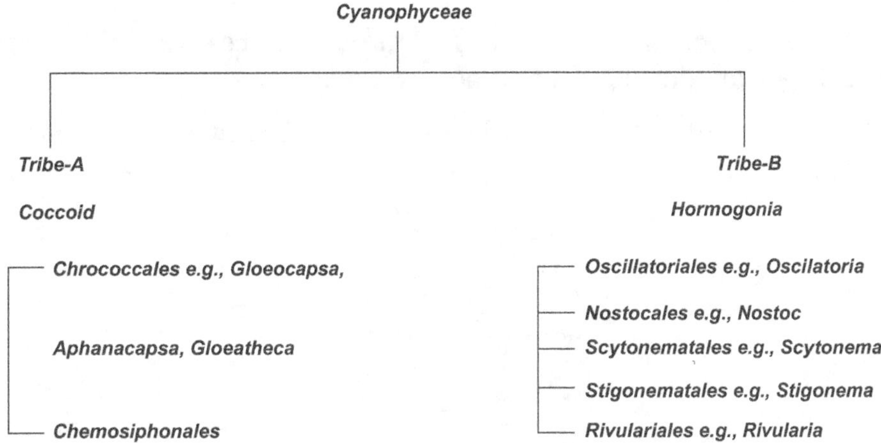

MICROCYSTIS

Systematic Position

Division   : Cyanophycophyta

Class   : Myxophyceae (or) Cyanophyceae

Order   : Chroococales

Family   : Chroococaceae

Genus   : *Microcystis*

Characteristic Features

1. *Microcystis* is a fresh water genus; they often cause water blooms, in hard water lakes.

2. The thallus is free-floating and varies in shape.

3. It contains a mass of single spherical cells but the sheaths of individual cells are confluent with colonial envelope.

4. The gelatanious matrix of the colonial envelope is of watery consistency and the margins of the matrix are usually not evident.

5. The cells frequently contain numerous pseudovacuoles.

6.   The reproduction of individual cells takes place by means of fission in three planes.

7.   The reproduction of the colony is through successive disintegration, each portion develops into a new colony.

8.   The shape of the colony is primarily determined by the environmental conditions.

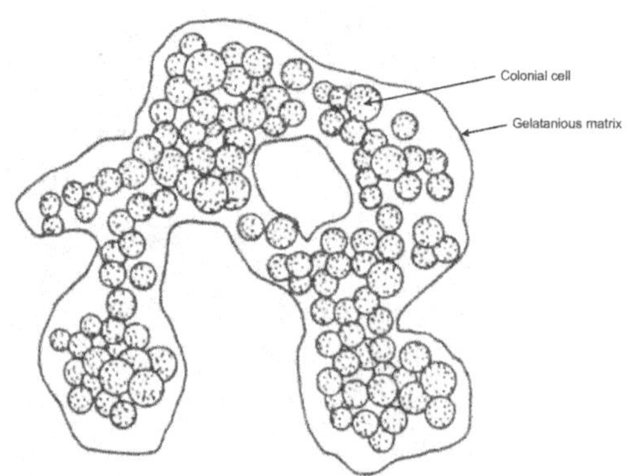

OSCILLATORIA

Systematic Position

Division : Cyanophycophyta

Class   : Myxophyceae (or) Cyanophyceae

Order   : Oscillatoriales

Family   : Oscillatoriaceae

Genus   : *Oscillatoria*

Characteristic Features

1.   It exhibits a very simple type of construction having unbranched trichomes.

2.   The trichome consists of a row of identical cells, and is surrounded by a sheath of mucilage.

3.   All the cells are capable of division and growth. In some species, the terminal cells are conical or rounded, often showing attenuation into capitate structure.

4. The intersurface of the terminal cell of the trichome may be covered by a thickened cap (calyptra) which is protective in function.

5. The reserve foods are cyanophycin granules and cyanophycean starch. They are found in the peripheral portion of cytoplasm.

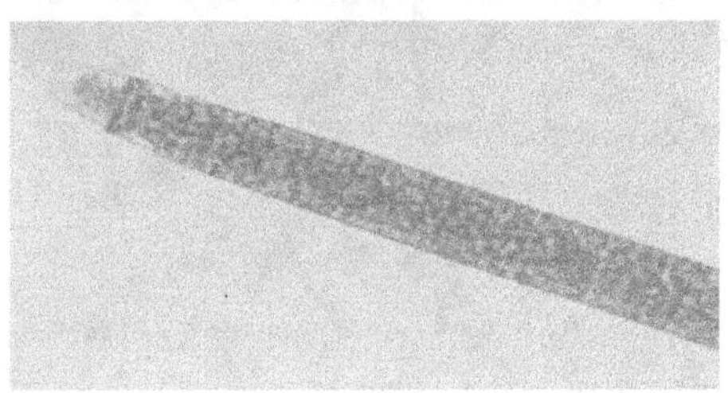

NOSTOC

Systematic Position

Division : Cyanophycophyta

Class   : Myxophyceae (or) Cyanophyceae

Order   : Nostocales

Family   : Nostacaceae

Genus   : *Nostoc*

Characteristic Features

1. *Nostoc* is a common blue green alga of filamentous form.

2. Species of *Nostoc* may be terrestrial or aquatic and generally occur in ponds, ditches and other pools of water as well as damp soil in the form of small, firm mosses of jelly.

3. A few species are endophytic in habit occurring in the intercellular cavities of plants like *Anthoceras,* Lemna and roots of *Cycas.*

4. Some lead a symbiotic life in association with a fungus-forming lichen.

5. Each gelatinous mass contains numerous unbranched filaments, which look very much like strings of beads under a microscope.

6. Each filament has a gelatinous sheath of its own, in addition to the gelatinous matrix of its colony.

7. The filament minus the sheath is commonly called a trichome.

8. Each cell of the filament is blue-green, more or less spherical or oval.

9. The characteristic feature of *Nostoc* is the presence of some enlarged vegetative cells, terminal or intercalary, with thickened walls and transparent content called heterocysts.

10. At the two poles of each heterocyst there are two pores, one at each end, through which cytoplasmic connection is maintained with adjacent cells.

11. The reproduction is vegetative and takes place by means of hormogones, akinetes and rarely heterocysts.

12. Under favourable conditions, the filament breaks into small pieces called hormogone, consisting of two or more cells capable of giving rise to a new plant.

13. During unfavourable conditions, the vegetative cells become thick walled, containing reserve food called akinetes or resting spore, which give rise to new filament during favourable conditions.

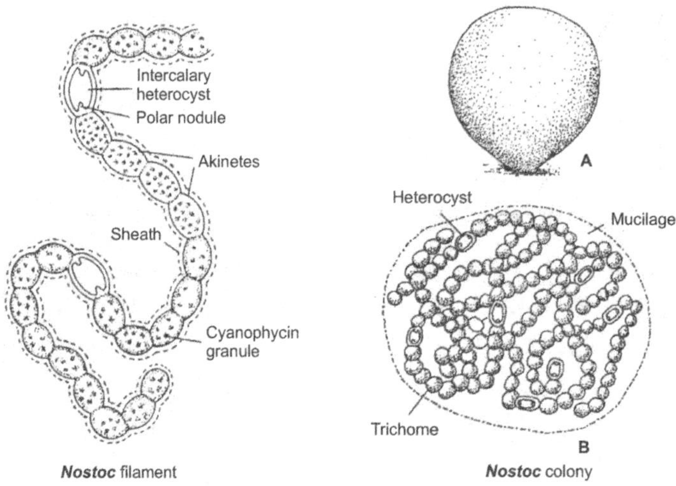

Nostoc filament   ***Nostoc*** colony

Spirulina

Systematic Position

Division : Cyanophycophyta

Class   : Cyanophyceae

Order   : Nostocales

Family   : Oscillatariaceae

Genus   : *Spirulina*

Characteristic Features

1. *Spirulina* is a multicellular freshwater blue green algae, commonly found in various polluted and unpolluted ponds.

2. The genus *Spirulina* contains the group of filamentous cyanobacteria characterized by spiral shaped chains of cell.

3. *Spirulina* shows the following characteristics: It has a trichome, spiriluna is deep blue green with regular spirals, 4-6 number per trichome, at the end conical, attenuated.

4. Trichome is 3 μm broad, 14.4 μm long, distance between two spirals being 31-33 μm.

5. The breadth of spirals is 27 μm. Cells have indistinct seta that are not clear.

6. Spirulina is a low fat, low calorie, cholesterol free source of easily digestible vegetable protein containing all the essential amino acids.

7. Spirulina has no cellulose in its cell walls.

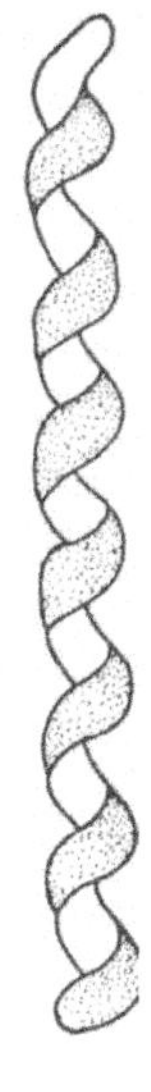

Anabaena

Systematic Position

Division : Cyanophycophyta

Class : Myxophyceae (or) Cyanophyceae

Order : Nostocales

Family : Nostacaceae

Genus : *Anabaena*

Characteristic Features

1. *Anabaena* is widely distributed and is commonly found in fresh water, floating freely or in thin mucous layers, and also often in wet soil rocks and tree trunks.

2. The plant body is unbranched, filamentous and consists of a row of oval or spherical cells. Bead like in appearance, intermixed at frequent intervals with some heterocysts and akinetes.

3. The sheath which surrounds the trichomes are hyaline and generally of watery nature.

4. The cells are usually, spherical or barrel-shaped, rarely cylindrical and never discoid.

5. The protoplast of the vegetative cell is usually filled with numerous pseudovacuoles.

6. The intercalary heterocysts are of the same shape as vegetative cells, though slightly larger ones are present in any trichome.

7. The reproduction is vegetative and takes place by means of hormogones, akinetes and rarely heterocyst.

8. Akinetes develop only next to heterocyst, always larger than the vegetative cells and generally cylindrical and with rounded ends.

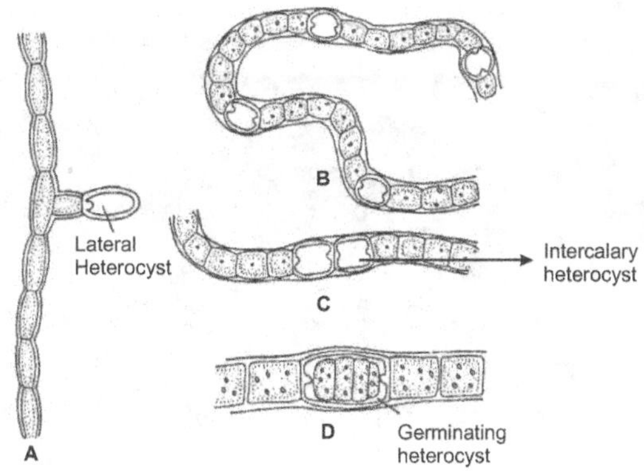

Lyngbya

Systematic Position

Division : Cyanophycophyta

Class   : Myxophyceae (or) Cyanophyceae

Order   : Nostocales

Family   : Oscillatoriaceae

Genus   : *Lyngbya*

Characteristic Features

1. The plant body is a uniseriate, unbranched filament. The trichomes occur singly in the enveloping sheath.

2. The members do not form spores or heterocysts. The reproduction takes place by vegetative method by means of hormogones.

3. The filament on breaking contains two or more cells called hormogones, each of which develops into a new plant.

4. Leathery prostrate mats on the substrate and very rarely false branched. Sheaths are always present and it is attached to the trichome.

5. Trichomes are cylindrical and may or may not be constricted at the cross walls.

6. Apical cells usually have a thickened outer wall or calyptra.

7. Reproduction occurs via trichome disintegration into usually short motile hormogonia, which often separate by necridia formation.

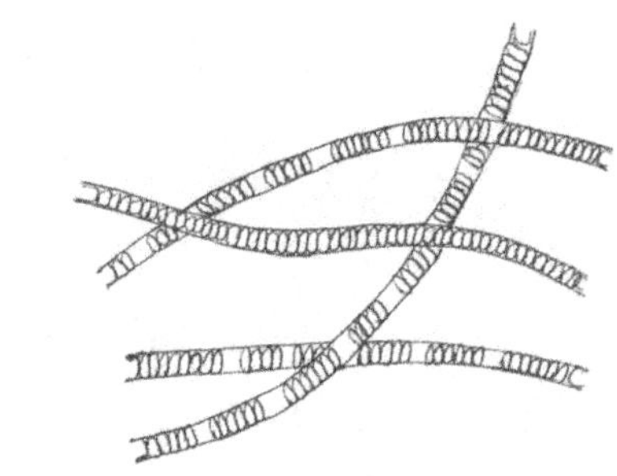

A portion with an enlarged view

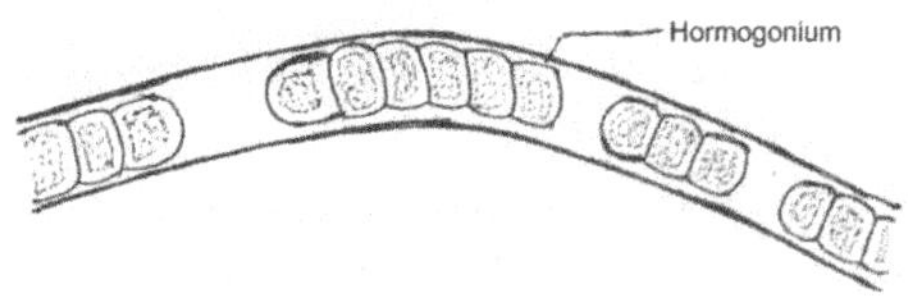

Hormogonium

Scytonema

Systematic Position

Division   : Cyanophycophyta

Class   : Myxophyceae (or) Cyanophyceae

Order   : Nostocales

Family   : Scytanemataceae

Genus   : *Scytonema*

Characteristic Features

1. The genus is usually found in subaerial habitats best on damp soil and rocky cliff.

2. The filament that consist of trichome and sheath posses distinct basal and apical regions forming little erect tufts.

3. The genus contains false type of branching.

4. The trichomes are usually of the same diameter throughout and with cylindrical cells.

5. The sheath which surrounds the trichomes is always firm, hyaline or coloured.

6. The sheaths are homogenous or lamellated.

7. The heterocysts are intercalary and are borne single or in twos or threes.

8. The heterocysts are of the same size as vegetative cell.

9. Akinetes are rare.

The genus contains false type of branching. The branches arise either between two heterocyst or else adjoining one as a result of the degeneration of an intercalary cell. The intercalary growth results in strong pressure being applied to the sheath, which finally ruptures and the trichome forms a loop outside. Further growth results in breaking of this loop and twin branches are produced. One or both of these branches may subsequently proceed to additional growth, the branch sheath extending back into parent sheath. Sometimes the false branching is initiated by degeneration of a heterocyst or a vegetative cell resulting in subsequent growth of two filaments on either side.

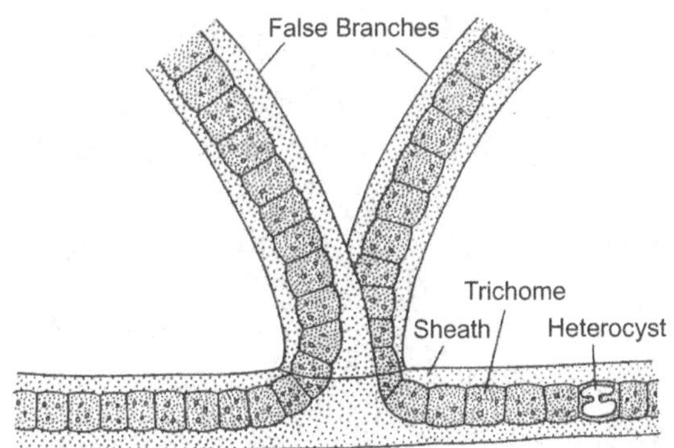

Stigonema

Systematic Position

Division : Cyanophycophyta

Class : Myxophyceae (or) Cyanophyceae

Order : Stigonemales

Family : Stigonemataceae

Genus : *Stigonema*

Characteristic Features

1. In Stigonema the threads consist of a single row or multiple rows of cells with definite apical growth.

2. Cells enclosed within a gelatinous sheath vary in thickness.

3. Behind the apical cell, one of the products divides twice to give a cruciform group.

4. One of these cells gradually moves into the centre and by further division a central cell surrounded by four pericentrals is produced.

5. These pericentrals divide again radially so that finally successive tires of pericentrals are produced, arranged approximately longitudinally. These cells are connected by protoplasmic strands.

6. The filaments are often aggregated to form a macroscopic yellow-green gelatinous mass.

7. The species are most common where there is continual dripping of water.

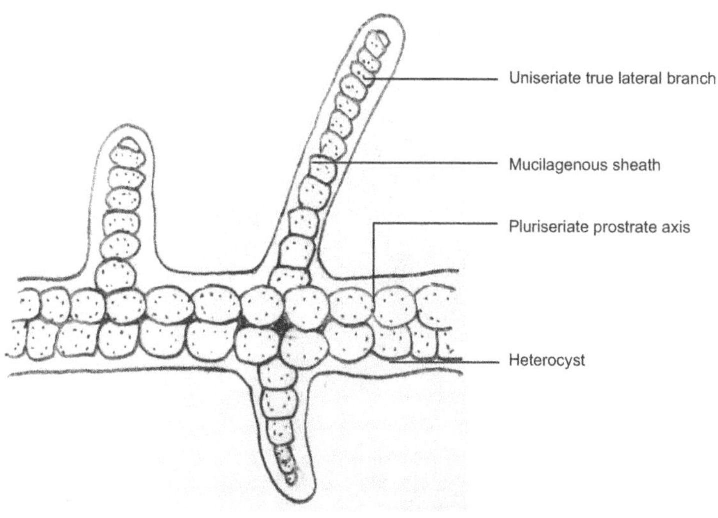

❧ APPENDIX- I ❧

SAFETY ON A COASTAL WALK

Points to Remember When you are on a Visit to the Coast

1. It is important to be aware of the state of the tide, and whether it is ebbing or flowing when you are on the shore. A set of tide table is very useful and costs little.

2. Estuary mud can be very sticky and it is best not to wander on to it; you may get trapped.

3. Many of our cliffs are made of crumbly rock, so do not go too close to the edge.

4. Always wear something on your feet when searching a rocky shore for sea creatures, branches especially have razor-sharp shells which easily cut the skin.

ROCKY COSTS

- With dozens of species of seaweeds and literally hundreds of different species of animals, including anemones, worms, molluscs, crustaceans, star fish, sea squirts and fish, beaches are a marvellous place to explore during ramble along the coasts.

- Not all beaches will turn up the same animals or plants for the degree of exposure to the waves has a definite effect upon the distribution of many organisms and very exposed areas of rock may be almost totally devoid of life. Also, soft rocks such as chalk and shale are too easily eroded by the sea and therefore support little life.

Once you are on the shore, the seaweeds are very obvious but many of the animals will have to be sought out under stones, amongst weed and in rock pods. There is, however, one most important rule to be followed when looking for crabs and other sea animals, it is that when you turn a stone over always turn it back again. Also when turning it back over be careful to make sure that you do not squash any of the small creatures that live under it; placing a small stone to lower a larger one down to create a gap beneath it will often prevent small creatures from being squashed, without having any detrimental effect upon this environment beneath the stone. The animal which live in sand are more difficult to find though a little digging with even a child's spade may turn up the odd cockle sand worm.

∾ **APPENDIX- II** ∾

PREPARATION OF CHEMICAL REAGENTS

Compensating for Hydrated States of a Chemical

If a method calls for an anhydrous chemical and the one available is a hydrated compound, one must compensate for the water. For example, if the method calls for 3 g of anhydrous sodium carbonate (Na_2CO_3, formula weight 106.00) and the reagent available is hydrated ($Na_2CO_3.10H_2O$, formula weight 286.16), one would divide the formula weight of the hydrated compound by the formula weight of the anhydrous compound and multiply by the amount of anhydrous compound called for in the method.

$$286.16 \text{ g} \div 106 \times 3 \text{ g} = 8.099 \text{ g of the hydrated compound}$$

The formula weight of water is 18.01

Dilutions

Preparation of Per cent Solutions from Solids (w/v)

Example

1% toluidine blue

Dissolve 1 g toluidine blue in sufficient solvent and make up to 100 ml.

Preparation of Per cent Solutions from Liquids (v/v)

Example

3% hydrochloric acid

Add 3 mL of concentrated hydrochloric acid to distilled water and make up to 100 mL.

Dilution of Already – Diluted Solutions

Example 1

5% hydrochloric acid from 36% hydrochloric acid: Pipette 5 mL of 36% hydrochloric acid into distilled water and make up to 36 mL. The percentage on hand (36) minus the percentage required (5) equals the volume of diluant to be added (31).

Example 2

100 mL of 30% alcohol from 95% alcohol

$$30 \times 100 \div 95 = 31.5 \text{ mL of 95\% alcohol required}$$

Percentage required $\times$ volume required $\div$ percentage on hand equals the amount of chemical on hand that must be diluted to the required volume. Although these are quick, practical examples, the general rule in all dilutions is $C_1V_1 = C_2V_2$ [concentration on hand (C_1) $\times$ volume (V_1) on hand $=$ concentration required (C_2) $\times$ volume (V_2) required].

Strengthening a Solution by Addition of a Solid

Example

To make 100 mL of a 25% solution from a 20% solution.

Take 90 mL (for example) of the 20% solution. It contains: $90 \div 100 \times 20 \text{ g} = 18 \text{ g}$

Add 7 g of the pure substance $(18 + 7 = 25)$ and dilute to 100 mL.

NORMAL SOLUTIONS

A normal solution is one gram molecular weight of the dissolved substance divided by the hydrogen equivalent (one gram equivalent) of the substance per liter of solution.

Example 1 (solids)

Normal sodium hydroxide (NaOH).

Relative molecular mass Na	=	23 valency
O	=	16
H	0 =	1
Total	=	40
Equivalent weight = 40 ÷ 1 =		40

$1N$ NaOH = 40 g NaOH dissolved in distilled water to a total volume of 1,000 mL.

Example 2 (liquids)

Normal hydrochloric acid (HCl)

The formula is: relative molecular mass ÷ (valency × specific gravity × concentration)

Relative molecular mass	=	36.4
Valency	=	1
Specific gravity	=	1.18
Concentration	=	36%
36.4 ÷ (1 × 1.18 × 0.36)	=	85.7 ml

1 N HCl = 85.7 mL concentrated HCl diluted to 1,000 mL.

Example 3 Normal sulfuric acid (H_2SO_4)

Relative molecular mass	=	98.08
Valency	=	2
Specific gravity	=	1.84
Concentration	=	98%
98.08 ÷ (2 ×× 1.835 ×× 0.98)	=	27.2 mL

1N H_2SO_4 = 27.2 mL concentrated H_2SO_4 diluted to 1,000 mL. All solutions are made up in distilled water unless the method states otherwise.

APPROXIMATE ATOMIC WEIGHTS OF SOME COMMON ELEMENTS

Hydrogen (H)	= 1	Magnesium	= 24
Carbon (C)	= 12	Phosphorus	= 31
Nitrogen (N)	= 14	Chlorine (Cl)	= 35.5
Oxygen (O)	= 16	Potassium (K)	= 39
Sodium (Na)	= 23	Calcium (Ca)	= 40

SOME USEFUL VALUES - pH

Hydrochloric acid 1 N	0.1
Hydrochloric acid 0.1 N	1.1
Hydrochloric acid 0.01 N	1.1
Acetic acid 1 N	2.0
Acetic acid 0.1 N	2.9
Sodium hydroxide 1 N	14.0
Sodium hydroxide 0.1 N	13.0
Potassium hydroxide 1 N	14.0
Potassium hydroxide 0.1 N	13.0
Potassium hydroxide 0.01 N	12.0
Calcium carbonate, saturated solution	9.4
Borax 0.1 N	9.2
Blood plasma (human)	7.3-7.5
Spinal fluid (human)	7.3-7.5
Gastric contents (human)	6.5-7.5
Duodenal contents (human)	1.0-3.0
Feces (human)	4.8-8.4
Urine (human)	4.8-8.4

GLOSSARY

Acronematic flagellum A flagellum without hairs.

Aerobic Needing oxygen.

Agar One or more polysaccharides containing sulfated galactose obtained from the walls of some red algae.

Agrophyte Red algae used in the production of agar.

Akinete Thick-walled resting spore.

Alginates Salts of polysaccharides composed of D-mannuronic and L-glucuronic acids obtained from brown algae

Alkalinity In water chemistry, the total quantity of base in equilibrium with carbonate or bicarbonate that can be determined by titration with strong acid. Alkaline waters have a high pH.

Allophycocyanin A blue biliprotein obtained from cyanobacteria and red algae.

Alpha granule Protoplasmic structure containing myxophycean starch in the cyanobacteria.

Amylopectin The storage polysaccharide in the cyanobacteria, composed of α-1,4-glucoside linkages, with 1,6 linked side chains.

Amyloplast A coloreless plastid containing starch.

Anaerobic Without oxygen

Anisogamy Fusion of gametes that are unequal in size or physiology

Anterior The front end, forward

Antheridium The sex organ in which the male gametes are formed

Antherozoid Male gamete

Anticlinal Perpendicular to the circumference of the thallus

Apical growth Growth by means of an apical cell dividing to form the thallus beneath it.

Aplanospore Colorless

Aerolate The chambers in the honeycomb arrangement of some diatom valves

Asexual Reproduction without the fusion of gametes

Autospore Aplanospore with the same shapes as the parent cell

Auxospore A resting cell in the diatoms that is commonly formed from the zygote.

Bacteriocin Antibiotic secreted by cyanobacteria that kills related strains of cyanobacteria.

Benthic Pertaining to any part of a lake or ocean bottom.

Benthos Organisms living on, and attached to, the bottom of aquatic habitats.

Biliprotein A red or blue pigment and attached bile (linear tetrapyrrole) chromophore in the cyanobacteria, cryptophytes, and red algae.

Biological carbon pump Trapping of carbon dioxide in the deep levels of the ocean.

Bioluminescence Emission of light by a living organism.

Biomass The amount of living organism at any one time in a habitat.

Bisexual Both sexes produced on the same individual.

Bisporangium A sporangium forming two meiospores in the red algae.

Bloom Heavy growth of planktonic algae in a body of water.

Brackish Saline water with a salinity less than that of sea water.

Calcareous ooze Bottom sediment in oceans composed of calcified remains of organisms.

Calcification Deposition of calcium carbonate, usually in association with smaller amounts of other carbonate.

Callose Polysaccharide associated with pores in sieve cells.

Carotene Oxygen-free, unsaturated, hydrocarbon carotenoid.

Carotenoid Yellow, orange, or red hydrocarbon fat soluble pigment.

Carpogonium Female gametangium in the red algae.

Carposporangium Carpospore-producing sporangium derived directly, or indirectly from the zygote in the red algae.

Carpospore Usually diploid spore produced by the carposporangium in the red algae.

Carposporophyte Usually diploid generation in the red algae derived from the zygote. It forms carpospores.

Carrageenan Red algal polysaccharide (polycocolloid) similar to agar; but needing higher concentrations to form a gel.

Cellulose Polysaccharide composed of β-1,4 linked glucose molecules that forms the main skeletal framework of most algal cells.

Cell wall A mostly rigid, often multilayered structure consisting of discrete microfibrillar polysaccharides embedded in an amorphous matrix composed of polysaccharides, lipid, and proteins, which together comprise an outermost layer of the cell proper.

Centric Type of ornamentation arranged at a central point in the diatoms.

Centriole Equivalent to a flagellar basal body.

Chitin Polysaccharide made up of repeating units of N-acetylglucosamine.

Chlorophyll Fat-soluble, green, porphyrin-type pigment.

Chromoplast or chromatophore A chloroplast with some other color than green.

Circadian rhythm Repeated sequence of events that occurs at about 24 hour intervals.

Coccoid Spherical

Coenobium Colony of algal cells in a specific arrangement and number that is fixed at the time of origin and is not subsequently augmented.

Coenocyte Large multinucleate cell without cross walls, except where reproductive bodies are concerned.

Colony A group of unicells which cohere and remain together as a unit.

Conjugation Fusion of two non-flagellated protoplasts.

Connecting band Part of the girdle in diatoms.

Contractile vacuole Vacuole fed by smaller vesicles that expels water and solute rhythmically outside the cell.

Cortex The outer portion of layer(s) of protist cells, including the plasma membrane, but excluding secreted non-living structures that may lie outside the plasma membrane. The outer portion of an algal thallus.

Cosmopolitan Occurring in many diverse places.

Cyanobacterocin An antibiotic produced by a cyanobacterium that inhibits growth of related cyanobacteria.

Cyanophage Virus in cells of the cyanobacteria.

Cyanophycin granule Polypeptide storage granule in the cyanobacteria.

Cyst A non-motile often dehydrated, resistant, inactive dormant stage in the life cycle.

Cystocarp In the red algae, the carposporophyte and surrounding gametophytic tissue (pericarp).

Cytokinesis Division of the cytoplasm usually right after karyokinesis (nuclear division).

Diatomaceous earth A mineral consisting of the remains of the silicified frustules of diatoms.

Dioecious An organism that has male and female gametes borne on separate plants.

Diploid Possessing two sets of chromosomes.

Ectoderm An outer protoplasm next to the plasma membrane in the cells.

Electron dense or electron opaque Term used to describe a material that absorbs electrons and appears dark in electron micrographs.

Electron transparent Term used to describe a material that does not absorb electrons and appears light in electron micrographs.

Endosymbiotic Term that describes an organism living inside a host in a symbiosis.

Epipelic Algae growing on the surface of sediments ranging from fine muds to coarse sands constitute epipelic algae.

Epiphyte One plant living on another plant.

Epitheca Large of the two halves of a diatom frustule.

Epizoic Living on an animal.

Epontic Living on the bottom of ice.

Estuary The mouth of a river where tidal effects are evident, and where freshwater and seawater mix.

Eyespot Red to orange area in a cell, composed of lipid droplets.

False branching In the cyanobacteria, breakage of a trichome through a sheath, giving the appearance of a branch.

Flagellate Unicell having at least one flagellum.

Floridean starch Red algal storage product composed of α-1,4 and β-1,6 linked glucose residues.

Floridoside Primary product of photosynthesis in the rhodophyta.

Fragmentation Type of asexual reproduction where a thallus breaks into two or more parts, each of which forms a new thallus.

Frustule Silicified cell wall in the diatoms.

Fucoidin Polysaccharide in the cell wall and mucilage of the brown algae composed of sulfated fucose units.

Gametogenesis The formation of gametes.

Gametophyte Plant generation that forms the gametes, usually haploid.

Gas vacuole A collection of gas vesicles.

Gas vesicle Hollow cylindrical gas-filled structures in the cyanophytea.

Generative auxillary cell Cell in the rhodophyta forming the gonimoblast filaments.

Geotaxis Movement of a cell away from (negative geotaxis) or towards (positive geotaxis) gravity.

Gliding Active movement of an organism in contact with a solid substrate where there is neither a visible organ responsible for the movement nor a distinct change in the shape of the organism.

Glycogen A storage polysaccharide related to amylopectin and stains red-purple with iodine.

Gonimoblast Usually diploid cells that form the carposporangia in the rhodophyta.

Haplobiontic Having only one multicellular stage.

Haploid Having one complete set of chromosomes.

Haplont An organism in which the only diploid stage is the zygote.

Hapteron or holdfast Bottom part of an alga that attaches the plant to the substrate.

Heterocyst Thick-walled, hollow looking enlarged cell in the cyanobacteria.

Heterotrichous Term used to describe division of a plant into an erect and a prostrate part.

Hydrophilic Water-attracting.

Hypotheca Smaller half of a diatom frustule.

Intercalary In between two cells or tissues.

Isogamy Fusion of similar gametes.

Isokont Cell with flagella of the same length.

Karyogamy Fusion of two gamete nuclei.

Kelp A member of the Laminariales (brown algae); also used for the burnt ash of plants of the Laminariales.

Laminarian Food storage polysaccharide in the brown algae composed principally of β-1,3 linked glucose residues

Lentic Related to a pond or lake.

Leucoplast Colorless plastid usually having a large number of starch grains and few thylakoids.

Lotic Related to rivers or streams.

Mucopeptide Polysaccharide of the walls of the cyanobacteria; composed of sugars and amino acids.

Multiseriate With more than one row of cells.

Mutualism A situation where two organisms benefit equally.

Myxophycean starch Storage polysaccharide of the cyanobaceria, similar to glycogen.

Niche The function of an organism within a community.

Nitrogen fixation The intracellular fixation of nitrogen gas from the atmosphere to ammonia in cyanobacteria.

Oogamy Fusion of a large non-motile egg with a small motile sperm.

Oogonium Single-cell female gametangium.

Oospore or zygospore Thick-walled zygote with food reserves.

Ostiole An opening to the outside in a conceptacle.

Ovum or egg Non-motile large female gamete.

Papilla A small rounded protuberance.

Paramylon Storage polysaccharide composed of β-1,3 linked glucose molecules.

Paraphysis Sterile structure found with sporangia or gametangia.

Parasite Heterotrophic organism that derives nutrients from a living host.

Pennate Term describing type of ornamentation.

Phycobiont Algal partner in lichen.

Phycocolloid Polysaccharide colloid formed by an alga.

Phycocyanin: Blue-green colored phycobiliprotein.

Phycoerythrin Pink-colored phytobiliprotein.

Physode Vesicle containing phaeophycean tannins in the brown algae.

Plastid Double membrane bounded organelle usually containing the photosynthetic apparatus or some part of it.

Plurilocular sporangium Many chambered sporangium in the brown algae, each chamber forming one swarmer.

Procarp Association of carpogonium and auxiliary cells in the rhodophyta.

Scytonemin Pigment that accumulate in the sheaths of cyanobacteria acting as a sunscreen to reduce the amount of near ultraviolet (370-384 mm) reaching the protoplasm.

Starch Storage polysaccharide composed of α,1-4 and α-1,6 linked glucose residues.

Stipe Organ between a holdfast and a blade.

Stria Row of punctuate (pores or loculi) in diatoms.

Supporting cell A cell that bears the carpogonial branch in some rhodophyta.

Tetrasporangium A sporangium producing four tetraspores, usually by meiosis.

Tetraspore Spore formed in a tetrasporangium, usually by meiosis.

Thylakoid Membrane-bound sac in a plastid.

Trichogyne Long colorless part of a carpognium that receives the spermatium in the Rhodphyceae.

Trichome A row of cells without the sheath in the cyanobacteria.

Uniaxial Having a main axis consisting of a single row of usually large cells.

Unilocular sporangium Sporangium composed of a single cell producing zoopores usually by meiosis.

Uniseriate Having a single row of cells.

Unisexual Having only one type of gametangium formed on one plant.

Upwelling An area of the ocean where nutrient rich bottom water rises to the surface.

Whiplash flagellum Flagellum without hairs on its surface.

Xanthophyll A carotenid composed of an oxygenated hydrocarbon.

Xylan Polysaccharide composed of xylose sugar residues.

Zoochlorellae Chlorophyta living inside invertebrate animals.

Zooplankton Animal plankton

Zoosporangium Sporangium that forms zoospores.

Zoospore Flagellated planospore.

Zygote Product of the fusion of two gametes.